English & Maths SATs Bumper Book Ages 10–11

2 in 1

Revision & Practice

KS2 Year 6

$a + b = c$

Perfect preparation for the Year 6 SATs

■SCHOLASTIC

First published in the UK by Scholastic, 2016; this edition published 2023

Scholastic Distribution Centre, Bosworth Avenue, Tournament Fields, Warwick, CV34 6UQ

Scholastic Ireland, 89E Lagan Road, Dublin Industrial Estate, Glasnevin, Dublin, D11 HP5F

www.scholastic.co.uk

A CIP catalogue record for this book is available from the British Library.

ISBN 978-0702-32678-3
Printed and bound by Bell and Bain Ltd, Glasgow

This book is made of materials from well-managed, FSC-certified forests and other controlled sources.

Due to the nature of the web we cannot guarantee the content or links of any site mentioned.

We strongly recommend that teachers check websites before using them in the classroom.

Every effort has been made to trace copyright holders for the works reproduced in this book, and the Publishers apologise for any inadvertent omissions.

Authors

Lesley and Graham Fletcher (English) and Paul Hollin (Maths)

Editorial team

Rachel Morgan, Vicki Yates, Audrey Stokes, Tracey Cowell, Rebecca Rothwell, Jane Jackson, Sally Rigg, Jenny Wilcox, Mark Walker, Red Door Media Ltd, Kate Baxter, Christine Vaughan and Julia Roberts

Design team

Dipa Mistry, Andrea Lewis, Nicolle Thomas, Neil Salt and Oxford Designers and Illustrators

Illustration

Judy Brown and Simon Walmesley

Contents

Maths Made Simple 91

How to use

This book has been written to help children reinforce the English and maths skills they have learned in school. Each subject is divided into sections covering a range of topics from the National Curriculum. Use the book little and often to practise skills and increase confidence. You can choose to work through the English and maths sections in order or focus on specific topics.

At the back of the book is a **Progress tracker** to enable you to record what has been practised and achieved.

English

<table>
<tr><td>

1 Chapter title

2 Topic title

3 Each page starts with a **recap** and a 'What is…' question which gives children a clear definition for the terminology used.

</td><td>

4 In the **revise** section there are clear explanations and examples, using clear illustrations and diagrams, where relevant.

5 **Tips** provide short and simple advice to aid understanding.

</td><td>

6 The **skills check** sections enable children to practise what they have learned with answers at the back of the book.

7 **Key words** that children need to know are displayed. Definitions for these words can be found in the **Glossary**.

</td></tr>
</table>

Maths

The Maths section has many of the same features of the English section and also some additional ones. Keep some blank or squared paper handy for notes and calculations!

2

Ratio and proportion: numbers

1

And, of course, three out of four of the squares are red.

3

↻ Recap

A fraction shows us one number compared to a whole. In the shape opposite, one out of four of the squares is blue.

Proportion is the fraction of a whole.
For this shape, the proportion of blue squares is one in four, or one out of four. And the proportion of red squares is three in four, or three out of four.

Ratio is different, because it compares amounts.
For the shape above, the ratio of blue squares to red squares is 1 to 3, or 1:3.

4

▤ Revise

Look at these examples.

In total there are 100 animals on a farm.
There are two dogs, three cats, five rabbits, 20 cows, 30 sheep and 40 chickens.

Proportion	Ratio
The proportion of dogs is two out of 100 animals. As a fraction this is $\frac{2}{100}$ or $\frac{1}{50}$.	The ratio of dogs to cows is 2:20. This can be simplified to 1:10. There are ten cows for every dog.
The proportion of rabbits is $\frac{5}{100}$ or $\frac{1}{20}$. One in every 20 animals is a rabbit.	The ratio of cows to chickens is 20:40. This can be simplified to 1:2. For every cow there are two chickens.

5

💡 Tips

- Proportion is a fraction of the whole; ratio compares different amounts.
- One in every five adults play computer games (so four out of five do not play).
  As a *proportion* this is one out of five, or $\frac{1}{5}$.
  But the *ratio* of adults who do play to adults who don't play computer games is 1:4.

6

💬 Talk maths

A wall is covered with 100 tiles.

Ten are black, 20 are white, 15 are red, 15 are yellow and 40 are blue.

Work with a friend to agree on some proportion and ratio statements about the tiles.

Remember to write the ratio in the simplest form.

7

✔ Check

1. Write the proportion of black squares in each pattern.

a. _____   b. _____   c. _____

2. Look at this pattern and write the ratios.

● ● ● ● ● ● ● ● ●

a. Blue to red ____:____   b. Red to green ____:____   c. Yellow to green ____:____

8

⚠ Problems

Brain-teaser In a class of 30 pupils, six of the class can speak two languages.

a. What proportion of the class can speak two languages? _____
b. What is the ratio of dual-language to single-language speakers? _____

Brain-buster A recipe for a fruit pie says to add blackberries and blueberries in the ratio 3:4.

a. If Hana has 15 blackberries, how many blueberries will she need? _____
b. What proportion of the berries will be blueberries? _____

132   133

---

1 Chapter title

2 Topic title

3 Each page starts a **recap** of basic facts of the mathematical area in focus.

4 In the **revise** section there are clear explanations and examples, using clear illustrations and diagrams, where relevant.

5 **Tips** provide short and simple advice to aid understanding.

6 **Talk maths** are focused activities that encourage verbal practice.

7 **Check** a focused range of questions, with answers at the end of the book.

8 **Problems** word problems requiring mathematics to be used in context.

Tips for using this book at home

Using this book, alongside the maths and English being done at school, can boost children's mastery of the concepts. Be sure not to get ahead of schoolwork or to confuse your child.

Keep sessions to an absolute maximum of 30 minutes. Even if children want to keep going, short amounts of focused study on a regular basis will help to sustain learning and enthusiasm in the long run.

Word lists These are the words you need to learn to spell.

Years 3–4

accident	certain	famous	island	peculiar	sentence
accidentally	circle	favourite	knowledge	perhaps	separate
actual	complete	February	learn	popular	special
actually	consider	forward/	length	position	straight
address	continue	forwards	library	possess	strange
answer	decide	fruit	material	possession	strength
appear	describe	grammar	medicine	possible	suppose
arrive	different	group	mention	potatoes	surprise
believe	difficult	guard	minute	pressure	therefore
bicycle	disappear	guide	natural	probably	though/
breath	early	heard	naughty	promise	although
breathe	earth	heart	notice	purpose	thought
build	eight/eighth	height	occasion	quarter	through
busy/business	enough	history	occasionally	question	various
calendar	exercise	imagine	often	recent	weight
caught	experience	increase	opposite	regular	woman/
centre	experiment	important	ordinary	reign	women
century	extreme	interest	particular	remember	

Years 5–6

accommodate	communicate	equip	immediately	physical	sincerely
accompany	community	equipped	individual	prejudice	soldier
according	competition	equipment	interfere	privilege	stomach
achieve	conscience	especially	interrupt	profession	sufficient
aggressive	conscious	exaggerate	language	programme	suggest
amateur	controversy	excellent	leisure	pronunciation	symbol
ancient	convenience	existence	lightning	queue	system
apparent	correspond	explanation	marvellous	recognise	temperature
appreciate	criticise	familiar	mischievous	recommend	thorough
attached	curiosity	foreign	muscle	relevant	twelfth
available	definite	forty	necessary	restaurant	variety
average	desperate	frequently	neighbour	rhyme	vegetable
awkward	determined	government	nuisance	rhythm	vehicle
bargain	develop	guarantee	occupy	sacrifice	yacht
bruise	dictionary	harass	occur	secretary	
category	disastrous	hindrance	opportunity	shoulder	
cemetery	embarrass	identity	parliament	signature	
committee	environment	immediate	persuade	sincere	

Multiplication table

x	1	2	3	4	5	6	7	8	9	10	11	12
1	1	2	3	4	5	6	7	8	9	10	11	12
2	2	4	6	8	10	12	14	16	18	20	22	24
3	3	6	9	12	15	18	21	24	27	30	33	36
4	4	8	12	16	20	24	28	32	36	40	44	48
5	5	10	15	20	25	30	35	40	45	50	55	60
6	6	12	18	24	30	36	42	48	54	60	66	72
7	7	14	21	28	35	42	49	56	63	70	77	84
8	8	16	24	32	40	48	56	64	72	80	88	96
9	9	18	27	36	45	54	63	72	81	90	99	108
10	10	20	30	40	50	60	70	80	90	100	110	120
11	11	22	33	44	55	66	77	88	99	110	121	132
12	12	24	36	48	60	72	84	96	108	120	132	144

English SATs Made Simple
Ages 10–11

Adjectives

What is an adjective?

↺ Recap

An **adjective** describes a characteristic of a noun.

What a **beautiful** day!

The word **beautiful** is an adjective.

▤ Revise

Adjectives describe or modify nouns. They give us more detail and precision in our writing.

an **arduous** journey ↑   a **wonderful** journey ↑

Not all adjectives describe characteristics we can see.
These adjectives describe very different journeys!

KEY WORDS

adjectives

✔ Skills check

1. Underline the adjectives in each sentence.

   a. The mischievous toddler hid in the large cupboard.

   b. It was a disastrous start to their annual holiday.

2. Replace the word 'nice' in these sentences with a more interesting adjective.

   a. Josh wrote a **nice** story.   Josh wrote a _____ story.

   b. Aliah enjoyed the **nice** pantomime.   Aliah enjoyed the _____ pantomime.

3. Add two adjectives to describe each noun.

   a. _____,

   _____ mountain

   b. _____,

   _____ bear

Nouns and noun phrases

↻ Recap

What is a noun?

A **noun** is a word for a person, place or thing. There are different types of noun: common and proper.

A **noun phrase** contains a noun as its main word and often contains a preposition or adjective: **next to the imposing school**.

What is a noun phrase?

📄 Revise

Common nouns

Names of things			
vehicle	people	month	business

Names of emotions and qualities			
happiness	joy	bravery	comfort

Proper nouns

Names of places, people, days and months			
Vietnam	Mr Jones	February	Wednesday

All proper nouns must start with a capital letter.

A **noun phrase** can be made by putting adjectives and nouns together:

long, hot month

Noun phrases may contain a noun and other words such as adjectives, determiners or prepositional phrases:

in the forest   at the park   many children

These are all noun phrases.

Adjectives + nouns = noun phrase
the amazing magical machine
Remember: a phrase does not contain a verb!

✔ Skills Check

1. Last <u>summer</u> we went on holiday to <u>Turkey</u>. Our <u>pleasure</u> was only curtailed when it was time to come home!

Tick the correct column to show the type of noun for each of the underlined words.

Word	Common noun	Proper noun
summer		
Turkey		
pleasure		

KEY WORDS

nouns
noun phrases

11

Verbs: present and past tense

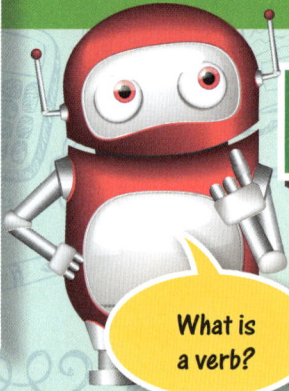

⟳ Recap

A **verb** tells you what is happening in a sentence. It is a doing word or being word.

The **tense** of a verb tells us when it happens: in the **present**, the **past** or the **future**.

What is a verb?

What is a tense?

KEY WORDS
verbs
tense (past, present)
progressive
future time

🗒 Revise

Action verbs ➡ I look   I wept

Simple present tense	Simple past tense
I look	I looked
he weeps	he wept
we drive	we drove

Use simple present or past for an action happening now (present) or an action that has already happened (past).

Being verbs ➡ I am   I have

Present progressive tense	Past progressive tense
we **are** revising	we **were** revising

Use a **helper verb** (**to be** or **to have**) to show the action is/was continuous.

Helper verbs: to be or to have. Not all past tenses end in ed!

✔ Skills Check

1. Complete the table using the correct form of the verbs.

Present tense	Past tense	Present progressive	Past progressive
she brings	she		
	they caught		
it grows		it is	
	we built		we

2. Complete the sentence using the past progressive form of the verb 'to work'.

They _____ hard when the fire alarm stopped them.

Verbs: present perfect and past perfect tense

↻ Recap

The **present perfect** is used for an action which happened at some time in the past.

> I **have been** to the theme park before.

The **past perfect** is used for something which happened before another action in the past.

> Jack **had never been** to a live football game, before Saturday.

What is the present perfect?

What is the past perfect?

Tips 💡

Perfect is a type of past tense. Don't be caught out by present perfect. It's still a past tense!

📋 Revise

Present perfect:
has/have + past tense of verb

They **have created** a beautiful picture.

↑ ↑

have + past tense of **create**
They **created** the picture.

She **has learnt** a new song.

↑ ↑

has + past tense of **learn**
She **learnt** a song in the past.

Past perfect:
had + past tense of verb

We **had enjoyed** the meal, until **the bill arrived**!

↑ ↑

action 1   action 2

I **had seen** the dark clouds before **the rain came**.

↑ ↑

action 1   action 2

KEY WORDS
present perfect
past perfect

✔ Skills Check

1. Complete the sentences using the present perfect of each verb in bold.

   a. **go**   He _____ out to play.

   b. **develop**   They _____ a method for baking perfect bread.

2. Rewrite this sentence using the past perfect for the verb in bold.

   I **enjoy** the film until the end spoilt it.

   _____

Verbs: tense consistency and Standard English

↻ Recap

What is tense consistency?

Tense consistency means having the same tense within a sentence.

Standard English is when the verb ending agrees with the thing or person doing the action. Standard English does not use slang or dialect words.

What is Standard English?

📄 Revise

Tense consistency: Use only one tense in a sentence:

They **booked** into their accommodation and **went** into the restaurant.

both verbs in past tense

Make the verb ending agree with the number of doers!

Standard English

A **singular** subject (or person doing it) must have a singular form of the verb:

Omar **visits** his grandparents. ➡ Omar **was visiting** his grandparents.

one person = singular form of verb

A **plural** subject (or things doing it) must have a plural form of the verb:

The children **clean** their teeth. ➡ The children **were cleaning** their teeth.

many people = plural form of verb

For collective nouns, one unit of people = singular form of the verb.

The team **is playing** in a local league.

KEY WORDS

singular
plural

✔ Skills Check

1. Rewrite this sentence in Standard English: I sees a bird in the garden.

_____

2. Choose the correct form of the verb to complete the sentence below.

Amy _____ Jack to the concert. **Tick one.**

is accompany ☐   have accompanied ☐   has accompanied ☐   has accompanies ☐

Modal verbs

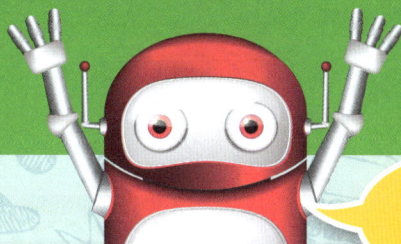

What is a modal verb?

↻ Recap

Modal verbs are **auxiliary verbs** that change the meaning of other verbs. The modal verbs are:

may   could   ought (to)   shall   will

might   should   would   can   must

Least likely ←————————————————→ **Most likely**

Modal verbs tell us how likely it is that something will happen.

Today is Monday so tomorrow **will** be Tuesday.

📋 Revise

Modal verbs tell us how likely an action is:

1. Whether someone is able to do something: Isaac **can** play guitar.

2. How likely something is: It **could** rain tomorrow.

They express degrees of certainty.

Must is more certain than **could**.   **Could** is less certain than **will**.

Learn these modal verbs:

We **must** be on time.   I **will** run quickly.   We **could** go swimming.

✔ Skills Check

1. **Underline the modal verbs in this sentence.**

   We could stay in on Saturday night but we might go to the cinema instead.

2. **Choose the best modal verb to fit in this sentence.**

   George _____ improve his backhand if he wants to win the tennis match.

3. **Which of these events is most likely to happen?**

   Tick one.   Emma will buy some jeans on Saturday.   ☐

   Emma should buy some jeans on Saturday.   ☐

   Emma ought to buy some jeans on Saturday.   ☐

KEY WORDS

auxiliary verbs
modal verbs

15

Adverbs

What is an adverb?

↻ Recap

An **adverb** describes a verb. It tells us how something is done.

📄 Revise

Adverbs give us more detail about a verb.
Adverbs often go next to the verb, but may go somewhere else in the sentence.

Adverbs describe the verb. They often end in **ly**.

It snowed. The sentence tells us it snowed, but not *how* it snowed!

| It snowed **soft**ly. | It snowed **heavi**ly. | It snowed **silent**ly. |

These **adverbs** describe the verb **snowed**.
Adverbs often end in **ly.** Each adverb changes *how* it snowed.

✔ Skills Check

1. **Circle the adverbs in each sentence.**

   a. He gently stroked the frightened kitten.

   b. They ran desperately to catch the train.

2. **Write a suitable adverb for each sentence.**

   a. She _____ opened the enormous parcel.

   b. We _____ searched the gloomy forest.

KEY WORD

adverbs

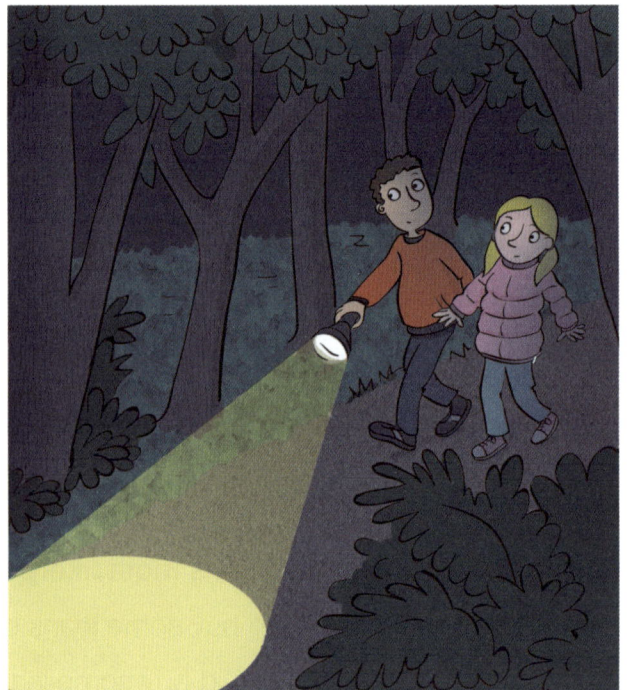

Adverbs and adverbials of time, place and manner

What is an adverbial phrase?

↻ Recap

An **adverbial phrase** tells us how, where or when something happened.

📄 Revise

Here are some examples of different types of adverb.
An adverbial phrase tells us:

- *how* it was done (manner) – **They climbed** with great determination.
- *where* it was done (place) – **The letter was posted** through the letter-box.
- *when* it was done (time) – **She read her book** before tea.

Be careful! Some words may be prepositions, such as **back** or **up**. Some words can also be **adjectives**, such as **slow**. Check how the word is used.

Adverbs of time	Adverbs of place	Adverbs of manner
during	above	some adverbs ending **ly**, such as angrily, carefully, accidentally
afterwards	abroad	wide
sometimes	behind	late
rarely	nowhere	fast
recently	west	slow
usually	indoors	well
frequently	towards	hard
today	over	
never	nearby	

KEY WORD

adverbials

✔ Skills Check

1. Tick the correct box to show the type of adverb for each word in bold.

Sentence	Adverb of time	Adverb of place	Adverb of manner
Rarely has the show been so successful.			
She practised **hard** for the piano test.			
They didn't know the treasure was **nearby**.			

Adverbs of possibility

↺ Recap

What is an adverb of possibility?

An adverb of possibility shows how certain we are about something.

These adverbs of possibility show we are sure of something happening:

definitely   **certainly**   **obviously**   **clearly**

These adverbs of possibility show we are less sure of something happening:

probably   **perhaps**   **maybe**   **possibly**

Revise

Maybe and **perhaps** usually come at the **beginning** of a sentence or clause.

Perhaps there will be ice cream for tea.
Maybe I can have a tablet for my birthday.

Other adverbs of possibility usually come in front of the main verb.

It is **clearly** going to rain.

However, they come after the verbs **am**, **is**, **are**, **was** and **were**.

It is **certainly** a busy road.

✔ Skills Check

1. **Choose the best adverb of possibility for each sentence.**

   a. It is _____ six miles to town.   **b.** I can _____ come to see you later.

   c. _____ we can have tea together?

2. **Explain how each adverb of possibility changes the meaning of the sentences below.**

   We are clearly going to win this game.   We are possibly going to win this game.

   _____

   _____

Fronted adverbials

↻ Recap

A **fronted adverbial** is an adverb or an adverbial phrase which is at the beginning of a sentence.

Fronted adverbials are usually followed by a comma.

📑 Revise

After the rain, we were delighted to see a magnificent rainbow.

↑ **Adverbial phrase** at the beginning of the sentence.
↑ rest of sentence
Adverbial of time describes *when* they can see the rainbow.

KEY WORD
fronted adverbials

Sadly, they packed their belongings and returned home.

↑ **Adverb of manner**, also a fronted adverbial.
↑ rest of sentence

In the water, they were surrounded by an abundance of coloured fish.

↑ **Adverbial of place** describes *where* there were fish.
↑ rest of sentence

A fronted adverbial comes at the front of the sentence. An adverbial phrase can come anywhere in the sentence.

✔ Skills Check

1. Rewrite the sentence below so it begins with a fronted adverbial. Use only the same words and remember to punctuate your sentence correctly.

The puppies played happily in the garden.

_____

2. Replace the fronted adverbial in this sentence.

Despite the traffic, they arrived at the party early.

_____

Clauses

↻ Recap

What is a clause?

A **clause** is a group of words which contain a subject (a person or thing who does the verb) and verb.

📄 Revise

A clause must have a subject and a verb. A clause makes sense, but may be dependent on other parts of the sentence.

The campsite was full.

A clause can be a complete sentence.
It has a subject (**the campsite**) and a verb (**was**).
A sentence can have more than one clause.

The campsite was full because it was a bank holiday.

first clause   conjunction   second clause

Often clauses aren't sentences, but they must have a subject and a verb.

in the garden   **they played** in the garden

no verb so not a clause   subject: they   verb: played – it's a clause!

✔ Skills Check

1. Tick the correct box to show if each group of words is a clause.

Group of words	Clause	Not a clause
they came home		
because they		
it was a wonderful beach holiday		

2. Underline the clause in each sentence.

  a. Despite the long delay, they arrived on time.

  b. They studied hard for their test.

KEY WORD
clauses

20

Main and subordinate clauses

↻ Recap

A **main clause** is an independent clause that makes sense by itself.

A **subordinate clause** is dependent on the main clause to make sense.

What are the different types of clause?

📄 Revise

Ranvir was late for school.

A **main clause** can be a complete sentence as long as it has a subject (**Ranvir**) and a verb (**was**).

Ranvir was late for school **because** the alarm didn't wake her.

main clause   conjunction   subordinate clause

                                   Tells us *why* Ranvir was late for school. Does not make sense by itself.

Even though it was very cold, **they went for a long walk.**

conjunction   subordinate clause   main clause
                 Can come first.   Does not have to come first.
                 Does not make   Makes sense by itself.
                 sense by itself.

KEY WORDS

main clause
subordinate clause

✔ Skills Check

1. Write a main clause for these sentences.

_____ in the evening.

_____ , although I can't make one.

2. Write a subordinate clause to complete these sentences.

a. I watched television until _____ .

b. We haven't got much bread though _____ .

Co-ordinating conjunctions

What is a co-ordinating conjunction?

↺ Recap

A **co-ordinating conjunction** joins two clauses which would make sense on their own.

📋 Revise

The co-ordinating conjunctions are:

| for | and | nor | but | or | yet | so |

An easy way to remember the co-ordinating conjunctions: the initial letters spell **fanboys**!

co-ordinating conjunction
↓
The magician waved his wand **but** the spell didn't work.
↖ ↗
Each part of the sentence makes sense by itself.

co-ordinating conjunction
↓
He spent several hours learning his spellings **yet** he didn't get them all right.
↖ ↗
Each part makes sense.

✔ Skills Check

KEY WORD
co-ordinating conjunctions

1. Choose the best conjunction for each sentence.

| nor | but | so |

a. My new bike was light _____ I was able to go very fast.

b. I like curry _____ I don't like it very spicy.

c. I wasn't able to score a goal _____ was I able to help my team score.

2. Join these sentences using the same conjunction.

a. I wanted a new tablet _____ they were very expensive.

b. The house was very cold _____ the central heating was on.

Subordinating conjunctions

What is a subordinating conjunction?

↻ Recap

A **subordinating conjunction** introduces a subordinate clause, which is dependent on the main clause.
Subordinating conjunctions include:

because   if   when   since

before   that   although   though

whenever   then   while   unless

📋 Revise

KEY WORDS
subordinating conjunctions

subordinating conjunction
↓
I can hold the dog while you bath it.
↑   ↑
main clause   subordinate clause

subordinating conjunction
↓
I have a snorkel although I don't know how to use it!
↑   ↑
main clause   subordinate clause

Remember, a subordinating conjunction introduces a subordinate clause.

✔ Skills Check

1. **Choose a different conjunction to introduce the final clause in each sentence.**

   a. I can't go swimming ＿＿＿＿＿＿ you give me a lift.

   b. I will go out with you ＿＿＿＿＿＿ you are free.

2. **Complete the sentences with a final clause.**

   a. More people came in after ＿＿＿＿＿＿＿＿＿＿.

   b. Even though you are my elder sister ＿＿＿＿＿＿＿＿＿＿.

Relative clauses

↺ Recap

A **relative clause** is a type of subordinate clause that adds information about a previous noun.

Relative clauses start with a **relative pronoun**:

that   which   who   whom

whose   where   when

Relative pronouns introduce a relative clause and are used to start a description about a noun.

What is a relative clause?

📄 Revise

KEY WORDS
relative clause
relative pronouns

The **man**, **whose car it was**, shouted angrily.

⬆
Relative clause, starts with **whose**.
Describes what the **man** owned. It modifies the noun.

The **lioness**, **which was only two years old**, was used to being with people.

⬆
Relative clause, starts with **which**.
Describes the **lioness**. It describes the noun.

Relative clauses are often enclosed by commas. They start with a relative pronoun.

✔ Skills Check

1. **Write a relative clause for these sentences.**

   a. The hotel, _____ , was next to the beach.

   b. August, _____ , is very busy.

2. **Put a tick to show the type of clause for the words in bold.**

Sentence	Main clause	Subordinate clause	Relative clause
The rain, **which fell heavily**, made us cancel the trip.			
We called at Tomas's house **after we had seen Josh.**			
Unless you are able to pay tomorrow, **the trip will be full.**			

Personal and possessive pronouns

↺ Recap

A **pronoun** replaces a noun. There are different types of pronoun. **Personal** and **possessive pronouns** are used to replace people or things.

What is a pronoun?

📋 Revise

The personal pronouns are: **I** **you** **she** **he** **it** **we** **they**

I wiped **the table** and put knives and forks on **it.**

The table is replaced by the **pronoun it** in the second clause.

Jamil and I were travelling by bus and **we** had a long journey.

Jamil and I is replaced by the pronoun **we** in the second clause.

KEY WORDS
pronouns
personal pronouns
possessive pronouns

There are also the possessive pronouns: **mine** **yours** **hers** **his** **its** **ours** **theirs**

These stickers are **mine.**

The **pronoun mine** is used to show possession of the stickers by **me.**

Using pronouns helps us to avoid repetition.

💡 Tips

male name → male pronoun	
he	his

female name → female pronoun	
she	hers

neutral (not male or female)	
it	its

plural names /objects → plural pronoun	
they	theirs

✔ Skills check

1. **Use the correct pronouns in each sentence.**

   a. Alicia enjoyed the party but _____ didn't like the food.

   b. George and Oscar went sledging which

   _____ **found enthralling.**

2. **Underline the word or words to which each pronoun in bold refers.**

   a. I have never used my fountain pen as **it** is too messy!

   b. John and I both devoured **our** food.

25

Prepositions

What is a preposition?

↺ Recap

A **preposition** links nouns, pronouns or a noun phrase to another word or phrase in the sentence.

📋 Revise

KEY WORD

prepositions

Here are some common prepositions:

about	above	across	after	around	as	at	before
behind	below	beneath	beside	between	by	for	from
in	in front of	inside	into	of	off	on	onto
out of	outside	over	past	under	up	upon	with

Prepositions often tell us the position of a person or object.

The **burglar** squeezed **between** the fence panels.

Preposition between describes the position of the **burglar**.

The **car** was **in front of** the garage.

Preposition in front of describes the position of the **car**.

JB 007

Prepositions can also be about time.

I went to the newsagents **before** school.
We don't finish school **until** 3.30.

✔ Skills Check

Prepositions should not go at the end of a sentence!

1. **Choose the best preposition for each sentence.**

   out of   without   around

   a. The hawk circled _____ its prey.   **b.** He took the milk _____ the fridge.

2. **Write a sentence using each preposition.**

   beneath   **across**

   _____

   _____

Determiners

What is a determiner?

↻ Recap

A **determiner** is used to define an object or person (a noun).

🗒 Revise

Let's look at the different types of determiner.

Articles	Quantifiers	Demonstratives	Possessives
the, a, an	All numbers: one, two... Ordinals: first, second... many, some, every, any	this, those, these	my, your, our, his her, their

These are just some examples – there are others.

I picked **some** apples from **the** tree and gave them to **my** mother.

quantifier   article   possessive

Each determiner defines the noun that follows it:
    some apples (not many or lots of)
    my mother (not anyone else's)

You don't need to know the names of each type of determiner but it might help to be aware of them.

✔ Skills Check

1. **Choose the best determiner for each sentence. Use each determiner once.**

    our   some   my   the

    a. I washed _____ face with _____ soap.

    b. We climbed up _____ stairs and reached _____ bedrooms.

2. **Circle each determiner in this sentence.**

    Every child must pay some money for the school trip.

KEY WORD
determiners

27

Subjects and objects

↻ Recap

What are subjects and objects?

Every sentence has a **subject**. The subject is the person or thing that does the action of the verb.

Many sentences have objects as well. The **object** has the action of the verb done to it.

Objects are usually nouns, pronouns or noun phrases.

📄 Revise

Let's look at an example.

I am selling.

This sentence has a subject but not an object.
The verb is selling. **I** am doing the selling. **I** is the **subject**.

I am selling **my bike**.

In this sentence:
I am still doing the selling so **I** is still the **subject**.
My bike is being sold. It is having the action of the verb done to it, so **my bike** is the **object**.

In a sentence:

- subjects usually come before the verb
- objects usually come after the verb.

Subject

FOR SALE

Object

KEY WORDS

subject
object

✔ Skills check

1. **Circle the subject in each of these sentences.**

   a. My mum drove the car.

   b. Our cat ate its food.

2. **Circle the object in each of these sentences.**

   a. Dad is making tea.

   b. The dog chased the cat.

Remember, subjects usually come before the verb. Objects usually come after the verb.

Active and passive verbs

What are active and passive verbs?

↻ Recap

Active and passive verbs are different forms of verbs.

📄 Revise

Most sentences use the **active** form of a verb. This means that the **subject** is doing the action and the **object** has the action done to it.

KEY WORDS
active voice
passive voice

Kelly scored all three **goals.**

Kelly has done the scoring so she is the **subject**.

The **goals** do not do the scoring. They have been scored so they are the **object**.

When the **passive** is used, the object moves to the front of the sentence and becomes the subject. The original subject moves to the end of the sentence but does not become the object. It becomes part of a **prepositional phrase**.

All three goals were scored by **Kelly.**

All three goals has moved to the front of the sentence and becomes the subject.

Kelly has moved to the end of the sentence.

The passive form can be used in formal writing.

To recognise the passive, look at the end of the sentence. It usually has **by someone** or **something** after the verb.

✔ Skills Check

1. Put a tick in the correct box to show whether each sentence is active or passive.

Sentence	Active	Passive
The winning shot was made by Alisha.		
The team won the league.		
Small mammals are hunted by eagles.		
Many people have climbed Mount Everest.		

2. Rewrite this sentence in the passive form.

The chef made a wonderful meal.

_____

Subjunctive

What is the subjunctive?

↻ Recap

The **subjunctive** is a form of verb used in formal speech or writing.

📄 Revise

The subjunctive uses only the simple form of a verb. For example, the simple form of **to run** is **run**.

The word **that** will help you to recognise the subjunctive. If the verb can be followed by **that** and something **should** happen, you will be using the subjunctive:

I demand that you be quiet.

Subjunctives are used in different ways:

- **verb + that**   to advise that   to ask that   to command that

  to demand that   to insist that   to propose that   to recommend that

  to request that   to suggest that

- **after phrases + that**   it is essential that   it is desirable that   it is vital that

- **I, he** or **she + were:** It is more natural to write **if I was to go to**, but this would be informal. The subjunctive form would be **if I were to go to**. This is known as the past subjunctive.

- **verb + that + be**   I **insist that** you **be** here.

✔ Skills Check

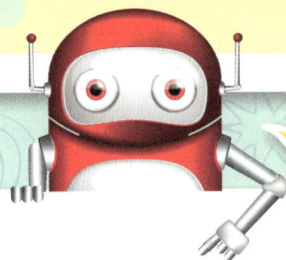

Subjunctives are only used in formal speech or writing. They are often used to suggest urgency or importance.

1. Add the subjunctive form of the verb in each sentence.

   a. It is important that you _____ on time for the show.

   b. If I _____ you, I would take the risk.

2. Underline the subjunctive in these sentences.

   a. If I were to give you £25, what would you do with it?

   b. The teacher asked that her students be quieter.

KEY WORD

subjunctive

Sentence types: statements and questions

What are the sentence types?

↺ Recap

There are four types of **sentence**: **statements**, **questions**, **exclamations** and **commands**.

KEY WORDS

sentence
statement
question
exclamation
command

▤ Revise

All sentences start with a capital letter.

A statement: states a fact and ends with a full stop.

> Dubai has a very busy airport.   My teacher, Mr Smith, is an accomplished pianist.

Both state a fact and end with a full stop = **statements**.

A question: asks a question and ends with a question mark.

> Where are you going?   Which event is most likely to happen?

Both ask a question and end with a question mark = **questions**.

💡 Tips

Questions often start with a question word:

who	why
what	which
where	when

They all start with **wh**!

Remember though that questions may start with other words.

✔ Skills Check

1. **Draw lines to join each sentence to the correct label.**

Sentence		Label
It is a sunny day.		statement
Is it sunny?		
What time does it start?		question
We can start it soon.		

2. **Write a question starting with the word below.**

When _____

31

Sentence types: exclamations and commands

📄 Revise

A command: tells someone to do something and can sometimes end with an exclamation mark.

It is sometimes called an imperative sentence.

No parking!   Line up!

Both are forceful **commands** and need an exclamation mark.

Please do not park here. ← This is not forceful. It is just a polite request. An exclamation mark is not needed.

An exclamation: expresses excitement, emotion or surprise and ends with an exclamation mark.

How marvellous!   What a terrifying tornado!

Expresses pleasure or excitement.   Expresses surprise or fear.
Both are **exclamations** and end with an exclamation mark.

Try saying a sentence. Think about what type of sentence it is. Are you asking a question? Do you need to sound forceful or surprised?

✔ Skills Check

1. Put a tick in the correct column to show the sentence type.

Sentence	Statement	Question	Exclamation	Command
Do you want a new bicycle				
Racing bikes are very aerodynamic				
What an amazing bicycle				
Ride this bike				

2. Write an exclamation starting with the word below.

How _____

Question tags

What are question tags?

↻ Recap

Question tags come at the end of a sentence. They try to make you agree with the sentence.

📄 Revise

Examples of question tags include: isn't it? don't you? wouldn't you?

They are called **question tags** because they are tagged onto the end of a sentence. They make statements into questions.

You all want to go on the trip.

↑ statement

You all want to go on the trip, **don't you?**

↗ The question tag makes this into a question.

Not all question tags are negative.

No one wants to miss the trip, **do they?**   means the same as   You all want to go on the trip, **don't you?**

↑ positive question tag   ↑ negative question tag

A question tag always comes after a comma.

✔ Skills Check

1. **Underline the question tags in these sentences.**

   a. You won't be late, will you?

   b. We're going to the cinema, aren't we?

2. **Add appropriate question tags to these sentences.**

   a. You'd like pizza for tea, _____ ?

   b. This is the right answer, _____ ?

Question tags are easy, aren't they?

Apostrophes: contraction

What is an apostrophe for contraction?

↻ Recap

An **apostrophe** for **contraction** is a punctuation mark used to show where letters have been missed out when two words are joined.

📋 Revise

We use the apostrophe to show where letters have been missed out.

I am = I'm
↑
missing letter – **a**

they could/they had = they'd
↑
missing letters – **coul/ha**

we are = we're
↑
missing letter – **a**

did not = didn't
↑
missing letter – **o**

The apostrophe must replace the missing letter or letters in the same place.

💡 Tip

Here are some common contractions:

you are	→	you're
there is	→	there's
was not	→	wasn't
could not	→	couldn't
have not	→	haven't
she will	→	she'll
could have	→	could've
they have	→	they've
I would	→	I'd
John is	→	John's

Exception to the rule:

will not	→	won't

When looking at a contraction, ask yourself which letters have been missed out. Where should the apostrophe go?

KEY WORDS

apostrophes
contraction

✓ Skills Check

1. **Circle the correct contraction in each sentence.**

   a. I wonder if **itl'l** / **it'll** / **i'tll** be sunny later.

   b. I **shouldve'** / **shouldv'e** / **should've** sent a birthday card to my gran.

2. **Write the correct contraction for these words.**

   a. had not _____

   b. could have _____

   c. we·would _____

Apostrophes: possession

What is an apostrophe for possession?

↻ Recap

An **apostrophe** and the letter **s** are often used to show **possession**; to show when an object belongs to someone or something.

📄 Revise

To use an apostrophe to show possession, you need to know if the possessor of the object is **singular** or **plural**. This will help you decide where to put the apostrophe.

KEY WORDS

apostrophe
possession
singular
plural

Single possessor

the car**'s** headlights

one car: **apostrophe + s**

the frog**'s** lilypad

one frog: **apostrophe + s**

this week**'s** work

one week: **apostrophe + s**

Plural possessors

the car**s'** headlights

several cars: **s + apostrophe**

the frog**s'** lilypad

several frogs: **s + apostrophe**

several week**s'** work

several weeks: **s + apostrophe**

Check how many possessors are there?
One possessor = apostrophe + s
Several possessors = s + apostrophe

💡 Tips

Before adding an apostrophe, be sure that you need to show possession.

The men enjoyed the game.

↖

Several men – no possession.

The men's game was enjoyable.

↖

The game belongs to the men – possession.

✔ Skills Check

1. **Rewrite each phrase using apostrophes to show possession.**

   a. The bags belong to the girls.

   _____

   b. The crayons belong to the boy.

   _____

2. **Insert an apostrophe into the correct place in the underlined word.**

   The <u>trains</u> arrivals were all delayed by the weather.

35

Commas in lists

How are commas used in lists?

↺ Recap

A **comma** is a punctuation mark that can be used to separate items in a list.

📄 Revise

Don't use a comma if there are only two items in the list.

I bought some bread and cheese.   two items: no comma needed

If you have more than two items in a list, use commas.

KEY WORD

comma

I bought some bread, cheese, grapes **and** chutney.

Use commas to separate each item.   Use **and** before the last item.

We went to Rome, Venice, Sorrento, Pisa **and** the Italian Lakes.

Use a comma after each item.   Do not use a comma before **and**.

✔ Skills Check

1. Insert commas in the correct places in these sentences.

   a. We had Jack Amir Rashid and Josef on our team.

   b. The children enjoyed their picnic of sausage rolls egg sandwiches apples crisps and juice.

2. Tick the sentence which uses commas correctly.

Europe is made up of many countries including Britain, France Spain Germany, and Italy. ☐

Europe is made up of many countries including Britain, France, Spain Germany and, Italy. ☐

Europe is made up of many countries including Britain, France, Spain, Germany and Italy. ☐

Commas to separate clauses

How do commas separate clauses?

↺ Recap

Commas can be used to divide clauses to make sentences easier to understand.

📄 Revise

Main clause and subordinate clause

I made tea **while** Asha set the table.

subordinating conjunction

subordinate clause at the end of the sentence: **no comma**

While Asha set the table, I made tea.

subordinating conjunction

subordinate clause at the start of the sentence: **comma always used**

Main clause and relative clause

relative pronoun

Cherry passed the ball to Donna, who scored easily.

relative clause: **comma usually used**

Clauses in the middle of a sentence

Sometimes the main clause is split by another clause.

main clause

Dev, **who is the best runner in the school,** won the county championship.

comma to separate clauses

relative clause

comma to separate clauses

✔ Skills Check

1. Insert commas into the correct places to separate the clauses.

   a. My favourite city is Paris which is the capital of France.

   b. Paris which is my favourite city is the capital of France.

Commas to clarify meaning

How are commas used to clarify meaning?

Commas are be placed in sentences to help us understand the meaning. Using commas within a sentence can help make the meaning clearer and avoid ambiguity.

📋 Revise

Sometimes the meaning isn't clear without commas.

In the following sentences the words are the same but the comma makes the meaning different:

"Let's eat Dad."

"Let's eat, Dad."

Someone is suggesting we should eat Dad!

That's clearer. Someone is telling Dad to eat.

The comma alters the meaning.

In the next two sentences, the commas alter the meaning again.

"My grandad in the distance could see a car." ← My grandad is in the distance and could see a car.

"My grandad, in the distance, could see a car." ← My grandad could see a car in the distance.

✔ Skills Check

1. Put commas in the correct places to make the meaning clear.

  a. My mum loves cooking my dad and me.

  b. Nate invited two boys John and Eddy.

  c. My uncle a singer and a dancer often appeared on television.

  d. Has the cat eaten Callum?

Commas after fronted adverbials

↺ Recap

What are commas after fronted adverbials?

A **fronted adverbial** is an adverb or an adverbial phrase, which is at the beginning of a sentence. A fronted adverbial is always followed by a comma.

Revise

At the end of the street, there is a large supermarket.

↗ fronted adverbial ↖ comma

Before going home, we made sure we had all of our belongings.

↗ fronted adverbial ↖ comma

Despite the price, we still bought a new fridge.

↗ fronted adverbial ↖ comma

KEY WORDS
fronted adverbials

✔ Skills Check

1. Place commas in the correct places in these sentences.

   a. Grim and sinister the graveyard lay before me.

   b. After lunch we had geography and art.

   c. Patiently I waited my turn.

   d. While the lead singer sang loudly the guitarist played the backing tune.

   e. Silently and softly the snow fell outside.

   f. Running to catch the bus I tripped and fell.

2. Write your own fronted adverbial for this sentence. Remember to punctuate it correctly.

   _____ it was a wonderful barbeque.

Inverted commas

What are inverted commas?

Inverted commas are also called 'speech marks'. They go around **direct speech** to show what is being said.

📋 Revise

Inverted commas go at the beginning and end of speech.

"Who said that?"

inverted commas   what is being said   **inverted commas**

Inverted commas *always* include one of the following punctuation marks:

| comma | full stop | question mark | exclamation mark |

The punctuation marks always come between the last word and the second set of inverted commas.

"I'm pretty sure that it was Christine!"

inverted commas   what is being said   inverted commas

punctuation

A **comma** is used when the writing continues past the end of the speech.

"You need to be able to use commas properly," said Mum. "They help you clarify meaning."

A **full stop** is only used when the speech is the end of the writing.
In this case, the comma moves in front of the first set of inverted commas.

Mum said, "You need to be able to use commas properly. They help you clarify meaning."

Question marks and **exclamation marks** are used in the same way depending upon the sentence types.

Mum said, "Why do you need to be able to use commas properly?"

Mum said, "You need to be able to use commas properly!"

KEY WORDS
direct speech
inverted commas

✔ Skills Check

1. Place inverted commas in the correct places in the following sentences.

   a. We can sit over there, said Demi.

   b. Sherri said, This punctuation stuff is easy.

2. Add the correct punctuation marks to the following sentences.

   a. Donny said " I like eating cream cakes "

   b. " They're not good for you " replied Shirley.

3. Rewrite the following sentences so that the words that are spoken come at the end of the sentences. An example has been done for you.

"We are going to the beach today," said Jerry.

Jerry said, "We are going to the beach today."

a. "On Sunday we are going to Nan's for tea," said my sister.

_____

b. "This bus is going to town," said the driver.

_____

💡 Tips

- Everything that is being said *and* a punctuation mark goes inside the inverted commas.
- Make sure you use the correct punctuation mark *before* the second set of inverted commas.

Remember the comma after words like **said** when you are using inverted commas.

Colons and semi-colons

↻ Recap

Colons (:) and semi-colons (;) are punctuation marks that are used within sentences to separate ideas.

Colons:
- introduce lists, summaries, examples and quotations
- mark the boundary between two independent clauses.

Semi-colons:
- show a link between two ideas
- separate complicated items within a list
- mark the boundary between two independent clauses.

Revise

Colons

Here are some examples of how colons introduce lists, summaries, examples and quotations.

List – To build this model tree you will need: glue, scissors, a ruler, tissue paper and some wire.

Summary – We have learnt the following: deciduous trees lose their leaves in winter.

Example – Examples of deciduous trees are: oak, sycamore, chestnut and poplar.

Quotation – Shakespeare wrote: "Under the greenwood tree, who loves to lie with me… Here shall he see no enemy but winter and rough weather."

Colons also mark the boundary between two independent clauses which make sense by themselves:

The weather was awful this weekend: it rained all Saturday and Sunday.

Semi-colons

Here is an example of how a semi-colon can show a link between two ideas.

Petrol is a fuel for cars; so is diesel.

Semi-colons also separate complicated items within a list.

I had to buy some garlic paste and tomatoes from the deli; some onions and potatoes from the grocer; and some plastic plates from the hardware shop.

As with colons, semi-colons mark the boundary between two independent clauses, but when the second part is related to the first part.

The weather was awful this weekend; I knew it would be.

✔ Skills Check

1. **Insert colons in the correct places.**

   a. You will need to bring with you your passport, plane tickets, money, sun cream and sunglasses.

   b. We now know some countries that border the Mediterranean Sea Egypt, France, Spain and Italy.

   c. Warm waters can be found in the Mediterranean Sea, the Caribbean Sea and the Indian Ocean.

   d. In *Julius Caesar*, Shakespeare wrote "There is a tide in the affairs of men. Which, taken at the flood, leads on to fortune."

   e. Phoebe often wears sunglasses the bright light hurts her eyes.

2. **Insert a semi-colon to link the two ideas.**

   May has thirty days so does June.

3. **Rewrite this bullet-point list as a sentence.**

   I need to go to
   - the supermarket for some dog food
   - the heel bar where my shoes have been mended
   - the library to get some books for my history project.

   Remember, colons and semi-colons are links. They go before the second part of a sentence.

   _____

   _____

   _____

4. **Insert a semi-colon to mark the boundary between the two independent clauses.**

   Water boils at 100 degrees Celsius at sea level it freezes at 0 degrees.

Parenthesis

↻ Recap

What is parenthesis?

Parenthesis is the term used for a word, clause or phrase that is inserted into a sentence to provide more detail.
- Parenthesis is what is written inside **brackets**.
- **Commas** and **dashes** can do the same job as brackets.

📋 Revise

Parenthesis does not make any difference to the understanding of the original sentence. It just gives the reader more information.

KEY WORDS

parenthesis
brackets
commas
dashes

The following sentence gives a piece of information:

> The Eiffel Tower is a very tall building.

By adding parenthesis, more detail is given but the meaning remains the same:

> The Eiffel Tower (which is in Paris) is a very tall building.
>
> ↑ ↑ ↑
> **parenthesis with brackets**

Commas and pairs of dashes can do the same job as brackets:

> The Eiffel Tower, which is in Paris, is a very tall building.
>
> ↑ ↑ ↑
> **parenthesis with commas**

> The Eiffel Tower – which is in Paris – is a very tall building.
>
> ↑ ↑ ↑
> **parenthesis with dashes**

Dashes tend to be used in less formal writing, such as in an email.

Remember, parenthesis is the information you add, not the punctuation around it.

✔ Skills Check

1. a. Insert the parenthesis into the following sentence, using brackets.

The three men talked quietly in the corner of the cafe. **Parenthesis:** *they looked like spies to me*

_____

b. Insert the parenthesis into the following sentence, using commas.

Denny is joining the army. **Parenthesis:** *my older brother*

_____

c. Insert the parenthesis into the following sentence, using dashes.

Suki won first prize at the dog show. **Parenthesis:** *a long-haired Alsatian*

_____

d. Insert the parenthesis into the following sentence, using either brackets, commas or dashes.

I had to keep very still while the doctor took my stitches out.
Parenthesis: *who was very gentle*

_____

e. Insert Parenthesis 1 and Parenthesis 2 into the correct places in the following sentence, using dashes and commas.

The strongest wind ever will hit this country.
Parenthesis 1: *A massive hurricane*
Parenthesis 2: *probably on Tuesday next week*

_____

_____

Hyphens

What is a hyphen?

↻ Recap

Hyphens are punctuation marks that are used to:
- join words together
- clarify meaning
- help pronunciation
- follow some prefixes.

Don't confuse hyphens and dashes. Dashes are longer and are used for parenthesis.

📄 Revise

Sometimes we join words together using a hyphen to show that they are linked.

> It was a **low-budget** film.

In this sentence, the film is neither low nor budget. We have to link the two words together to get low-budget, meaning it did not cost much.

The meaning of some sentences isn't clear without a hyphen.

> Joe Montana was a famous American football player.

The sentence is ambiguous. Was Joe a famous American who played football, or was he famous for playing American football? Adding a hyphen shows that Joe played American football.

> Joe Montana was a famous **American-football** player.

Without a hyphen, we would not know how to pronounce words like **re-enter**. The hyphen tells us that the letters on either side of it are both pronounced.

Here are some examples of when hyphens follow prefixes.

ex-police officer   **all-**inclusive   **self-**conscious

✔ Skills Check

KEY WORD

hyphen

1. **Insert hyphens to join the correct words together.**

   My mother in law is coming for Sunday lunch.

2. **Insert a hyphen in this sentence to make it clear that the instruments have not been used much.**

   My uncle, a retired surgeon, showed me some of his little used instruments.

3. **Rewrite 'resign' with a hyphen to show that it means 'to sign again'.**

   _____

Ellipsis

What is ellipsis?

↺ Recap

Ellipsis is the omission of repeated, predictable or unnecessary words.

🗐 Revise

Imagine this question has been asked: "Where are you going?"

This is the reply: "I'm going to the skate park."

Not every word of the reply is necessary. It could just have been "To the skate park." or even "The skate park."

Sometimes words are repeated in a response. "Who was the prime minster in 1943? "The prime minster in 1943 was Winston Churchill." could just have been "Winston Churchill."

In this sentence, 'some' has been repeated when it doesn't need to be.

We have bought some apples, some oranges, some bananas and some pears.

The sentence could be: We have bought some apples, oranges, bananas and pears.

✔ Skills Check

KEY WORD
ellipsis

1. Underline the unnecessary words.

a. My son was born at the start of this century, in 2001.

b. I went because I wanted to go. c. My sister likes salad but I don't like salad.

2. Insert the words that are missing.

a. "Do you want to go to the park?" "_____ we don't."

b. "Have you finished your homework?" "_____ I have.

c. "Who does the cooking in your house?" "My dad _____."

47

Paragraphs

What are paragraphs?

↺ Recap

Paragraphs organise writing to make it easier to understand.
- They break text down into small sections so it is easy to read.
- They are a series of sentences about the same idea.
- We start a new paragraph for each different idea, place, time, character or event.

Revise

In the following story, Arty is a police officer.

A shiver ran down Arty's spine. Night observations. She hated them. She was always scared because she never knew how they would turn out.

new paragraph for a new idea

She heard footsteps approaching from her right-hand side. It was Ben, Arty's work partner, who had been checking out the other end of the street.

new paragraph   different idea   different event

It was time to move. Together they crept from their hiding place towards the warehouse, which lay before them, dim, dark and threatening.

new paragraph   different place   different character

Inside the warehouse, Jim Evans and three of his gang were loading up a van.

✔ Skills Check

1. Draw two lines (//) to show where a new paragraph should begin. Give a reason for your answer.

Jim Evans was a small-time crook who made his money by selling stolen goods. He had often been in trouble with the law. Outside the warehouse, Arty and Ben waited patiently for back-up. They knew they couldn't do this alone.

Reason: _____

Headings and subheadings

What are headings and subheadings?

↺ Recap

Headings are titles for pieces of writing – they go at the start of the piece.

Subheadings are titles for sections of writing within a longer piece – they go at the start of the section.
- They make the writing easier to read by structuring it.
- They often summarise the writing.

📄 Revise

Surfing! ⟵ **Heading** – tells us what the whole piece is about.

Basic equipment ⟵ **Subheading** – gives a summary of this section.

It's exciting and dangerous but surfing is growing in popularity. If you want to try it, there are two things you absolutely have to have: a surf board and big waves. You can do something about the board but you can't do much about the waves!

💡 Tips

The text in a question will normally be more than one paragraph long. **Read all of it** and decide what the **main idea** is. That will be the **heading**. Then try to give **short summaries** of **each section**. These will be the **subheadings**.

✔ Skills check

1. **Read the following text and suggest a heading and a subheading.**

   a. Heading: _____

   Did you know that not everyone speaks like you? I don't mean people in other countries with other languages. I mean here, in Britain. It's really strange, some words are the same but they aren't pronounced anything like each other.

   b. Subheading: _____

   My cousins, who live a hundred miles away, say "Barth", while I know it should be pronounced "Bath". They say, "scon". Don't they know it's "scone"? They don't say "ston" when they mean "stone".

Synonyms

What are synonyms?

↻ Recap

Synonyms are words with the same or similar meaning.

📋 Revise

Using different synonyms for words can make our writing more interesting.

It is a **big** elephant.

Large, **enormous** and **massive** are all synonyms for big.

"That is an **enormous** elephant," **said** Ranvir.

The word **said** can be changed for a more interesting synonym:

declared spoke uttered pronounced

enormous

BIG

large

massive

💡 Tips

When writing, ask these questions.
- What other words mean the same?
- Are they more interesting or precise?

Can you think of some synonyms?

KEY WORD

synonyms

✔ Skills check

1. Tick all the synonyms for the word 'difficult'.

complex ☐

arduous ☐

effortless ☐

intricate ☐

easy ☐

2. Draw lines to match each word to its synonym.

ancient	genuine
curious	antique
familiar	inquisitive
sincere	known

Antonyms

What is an antonym?

↻ Recap

Antonyms are words with the opposite meaning.

Revise

Light is the antonym of heavy. It has the opposite meaning.

backward ⟷ forward

Moving **forward** is the opposite of moving **backward**. **Backward** is the antonym of **forward**.

Here are some more examples of antonyms.

Word	Antonym
encourage	discourage
guilty	innocent
night	day
singular	plural

Tips

Sometimes adding a prefix to a word can create an antonym!
- happy ➔ **un**happy
- encourage ➔ **dis**courage

✔ Skills Check

KEY WORD

antonyms

1. **Draw lines to join each word to its antonym.**

healthy   minimum

young   mature

permanent   unwell

maximum   temporary

2. **Choose an antonym to replace each word in bold.**

a. I **made** a massive tower. _____

b. The successful man was very **humble**. _____

c. The **foolish** child had no packed lunch. _____

Prefixes: in or im? il or ir?

What is a prefix?

↺ Recap

A **prefix** is added to the beginning of a word to change it into another word, with a different meaning.

🗒 Revise

Each prefix has a different meaning.

The prefixes **in**, **im**, **il** and **ir** all have negative meanings (they often mean **not**).

inaccurate   **incredible**
↑   ↑
not accurate   **not** credible

im is used with words beginning with **p** or **m**

impossible   **im**mobile
↑   ↑
not possible   **not** mobile

il goes before words beginning with **l**

illegal   **il**logical
↑   ↑
not legal   **not** logical

ir goes before words beginning with **r**

irresponsible   **ir**retrievable
↑   ↑
not responsible   **not** retrievable

✔ Skills Check

1. Choose the correct prefix to change each word to its opposite meaning.

ir   il   im   in

a. ____mature

b. ____relevant

c. ____accessible

d. ____legible

Tips

Think about the letter that the root word starts with.
- **il** = words starting with **l**
- **ir** = words starting with **r**
- **im** = words starting with **p** or **m**

For all other words, try the prefixes **un**, **in**, **dis** or **re**.

KEY WORD

prefix

2. Circle the correct use of a prefix in each sentence.

a. Emily had several **imcorrect** / **ilcorrect** / **incorrect** answers.

b. They waited **impatiently** / **inpatiently** / **irpatiently** to be chosen for the team.

c. An **ilregular** / **inregular** / **irregular** hexagon has six unequal sides.

Prefixes: re, dis or mis?

📄 Revise

The prefix **re** means **again** or **back**. It changes the meaning of the word.

regain
↑
to gain **again**

resolve
↑
to solve **again**

readjust
↑
to adjust **again**

The prefix **dis** changes the verb to its opposite meaning (often means **not**).

disable
↑
not able

disinterest
↑
not interested

disbelief
↑
not believing

The prefix **mis** also changes the verb to its opposite meaning (often to **do it badly**).

misuse
↑
to use **badly**

mistreat
↑
to treat **badly**

misbehave
↑
to behave **badly**

✔ Skills Check

1. Draw lines to match the prefixes to the root words. Then write each new word.

re   loyal   _____

dis   judge   _____

mis   design   _____

2. Use a prefix to change the meaning of these words so they match their definitions.

Word	New word	Definition
place		to put back again
calculate		to work out wrongly
tasteful		objectionable

Suffixes: ous, cious or tious?

What is a suffix?

↻ Recap

A **suffix** is used at the end of a word, to change it into another word and to change its meaning.

📝 Revise

If words end in **ce**, change the **ce** to **cious**.

vice → vi**cious**   grace → gra**cious**

If words end in **tion**, change the **tion** to **tious**.

cau**tion** → cau**tious**   ambi**tion** → ambi**tious**

There are exceptions to this rule. For instance, fic**tion** → ficti**tious**

Add the suffix **ous** to words ending in a **consonant**.

danger → danger**ous**   poison → poiso**nous**

If words end in **our**, change **our** to **or** and then add **ous**.

hum**our** → hum**orous**   glam**our** → glam**orous**

Words ending in **ge** keep **ge + ous**.

coura**ge** → coura**geous**   advanta**ge** → advanta**geous**

✔ Skills Check

KEY WORD
suffix

1. Add the correct suffix to make a new word. Write the word.

ous   cious   tious

a. malice _____   c. vigour_____

b. infection _____   d. mountain _____

2. Circle the correct spelling of each word.

a. religous / religious / religeous   b. consciencious / conscientious / consciencous

Suffixes: ant or ent? ance or ence? ancy or ency?

📄 Revise

To work out which suffix to use it helps to know that some of them are related.

Words ending in **ation** often use the **ant**, **ance** or **ancy** suffixes.

Let's look at some:

hesit**ation** → hesit**ant** → hesit**ance** → hesit**ancy**

They all have **a** in the suffix.

Use **ent**, **ence** or **ency** after a soft **c** sound or after **qu**:

inno**cent** → inno**cence**   fre**quent** → fre**quence** → fre**quency**

Tips 💡

There are words that don't follow these guidelines, which you need to learn.

For example: independ**ent**   assist**ance**

If one of a word's suffixes has an **a** in it, others might: assist**ant**   assist**ance**

✔ Skills Check

1. **Choose the correct suffix for each word. Then write out the words in full in the boxes. In some cases none of the suffixes make a word, so some rows should be left blank.**

Start of word	ant or ent?	ance or ence?	ancy or ency?
dec			
confid			
toler			
obedi			

2. **Circle the correct spelling to complete each sentence.**

   a. The **non-existance / non-existence** of dodos in Mauritius has long been a cause for regret.

   b. Your help is more of a **hindrance / hindrence**.

   c. Please complete the **relevent / relevant** application form.

Word families

What is a word family?

↻ Recap

📄 Revise

Adding a prefix or suffix will change the meaning of the word and might change its function.

Start with a **root word** and then try adding different prefixes and suffixes. How has the word changed?

unattach ← **attach** → attach**ment**
attach**ed**

These words all have the root word **attach** but have different beginnings and endings.

indefinite ← **definite** → definite**ly**
defini**tion**

These words all have the root word **definite** but have different beginnings and endings.

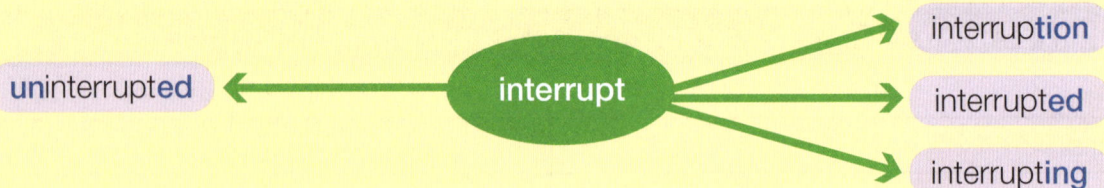

uninterrupt**ed** ← **interrupt** → interrup**tion**
interrupt**ed**
interrupt**ing**

These words all have the root word **interrupt** but have different beginnings and endings.

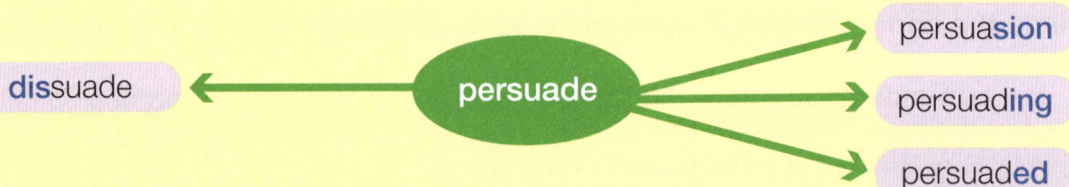

dissuade ← **persuade** → persua**sion**
persuad**ing**
persuad**ed**

These words all have the root word **persuade** but have different beginnings and endings.

✔ Skills check

1. Make word families by adding two suffixes to each root word. Write the new words in full.

   a. critic _____ _____

   b. depend _____ _____

   c. achieve _____ _____

KEY WORDS

word families
root word

2. Split each word into its different parts.

Word	Root word	Suffix	Prefix
impatience			
unfriendly			
disappointment			

3. Underline the word which does not belong to each word family.

   a. unbelievable, disbelief, disembark, believed

   b. redecorate, indecent, decoration, decorated

   c. immaterial, materially, materialise, maternal

If you know a root word and can spell it, you can then make lots of other words, using prefixes and suffixes.

4. Write the root word used in each of these words.

   a. disenchantment _____

   b. unenthralling _____

   c. misapplication _____

Always look for the common features to find a word family.

Tips 💡

Sometimes when you add a prefix or suffix to a root word, you need to lose some letters first.
- For words ending in **e**, lose the final **e**.
- Persua**de** + sion: lose **de** and add suffix.

Letter strings: ought

↻ Recap

What is a letter string?

A letter string is a group of letters which make one sound, within a word.

The letters **ought** can be used to make many different sounds.

📄 Revise

These are the most common **ought** words:

ought	thought	bought	brought
sought	fought	nought	wrought

This letter string makes an **ort** sound.

However, in dr**ought** these letters make an **out** sound (as in shout).

Learn the letter string ought and spelling these words will be easy.

✔ Skills Check

1. Choose the best word to go in each sentence.

brought   bought   sought   wrought

a. They _____ a way out of the forest, but it was hard to find.

b. I _____ some toys with me.

c. They installed a new _____ iron gate.

d. We _____ some cakes to have with our sandwiches.

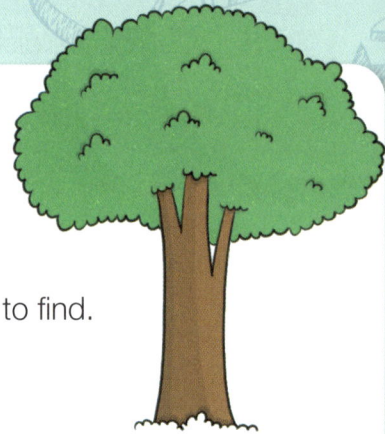

2. Draw lines to match these words to their definitions.

ought	nothing
fought	considered
nought	struggled
thought	should

💡 Tips

Be careful: some words have the same sound but are spelled differently, for example **caught** and **court**, **taught** and **taut**.

Letter strings: ough

What sound does the letter string ough make?

↻ Recap

The letter string **ough** makes several sounds: **uff** (as in stuff); **off**; **oo** (as in moon); **oe** (as in toe); and **ow** (as in cow).

Revise

Using the letter string **ough** can be tricky because it can make so many sounds. Here are some examples of each sound.

uff (as in cuff)	off	oo (as in moon)	oe (as in toe)	ow (as in cow)
rough	cough	through	though	bough
tough	trough		dough	plough

Some **ough** words don't belong in these groups:

thorough   **borough**

These words both have an **uh** sound at the end.

💡 Tips

Say the word, then work out which sound it makes.

✔ Skills Check

1. Use these 'ough' words to make new words to fit in each space.

   dough   **rough**   **tough**

   a. They _____ worked out how to make the model.

   b. The doors were made of _____ glass.

   c. We bought some _____ to eat at the fairground.

2. Write the 'ough' words to match each definition.

Definition	Word
area	
branch	
cultivate	
sufficient	
animal food container	

Silent letters

When are silent letters used?

↻ Recap

Silent letters are used to write a sound – but you can't hear them when you say the word.

目 Revise

There are lots of silent letters. They often pair up with another letter:

bt has a silent b

doubt	debt	subtle

you only hear the **t** sound

mn has a silent n

solemn	column	autumn

you only hear the **m** sound

s can be a silent s

island	aisle	debris

you cannot hear the **s** sound

kn has a silent k

knight	knowledge	knit

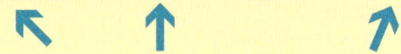

you only hear the **n** sound

st has a silent t

thistle	whistle	castle

you only hear the **s** sound

✔ Skills Check

1. Underline the silent letter in each word in bold.

   a. They went over the bridge to the **Isle** of Anglesey.

   b. Look at the third **column**.

   c. We are **indebted** to you, thanks to all your efforts.

   d. Caitlin saw a **Mistle** Thrush in the garden.

2. Circle the correct spelling for each word.

   a. dought / doubt / dout

   c. condam / condemn / condem

   b. isle / iall / iel

   d. brissle / brissel / bristle

💡 Tips

To help you spell a word, pronounce it with the silent letter: sub–tle.
If you can hear each letter, you will use it when writing the word.

c or s?

What kind of sound can c and s both make?

c and s can both be used to make a soft **s** sound (as in sun).

📋 Revise

How do we know when to use **c** or **s** at the start of a word?

When the next letter is a consonant we must use **s**:

    scrap   **s**mell   **s**nore

When the next letter is **a**, **o** or **u** we must use **s**:

    sanity   **s**ock   **S**unday

When the next letter is **e**, **i** or **y** we use **s** or **c**:

    seven   **s**ingle   **s**ynonym

    cement   **c**igar   **c**ygnet

💡 Tips

Look out for common word endings using the soft **s** sound:

nce → fe**nce**
nce → adva**nce**

rce → pie**rce**
rce → resou**rce**

After a short vowel sound in short words we use **ss**:   k**iss**   m**iss**   ch**ess**

In longer words we use **ice**:   prejud**ice**   precip**ice**   off**ice**

    f**ace**   sp**ace**   r**ace**   all use **ace**.

✔ Skills check

1. Draw a circle round the correct spelling of the words in bold.

a. I went **twice / twise** to call on Ahmed.

b. There was a very **fierse / fierce** dog behind the gate.

c. It was **bliss / blice** sitting in the hot sun.

d. Our teacher **cuggested / suggested** that we read books by Michael Morpurgo.

2. Write the correct spelling of the word highlighted in bold.

a. Our doctor's **practiss** is very busy. _____

b. I asked for a **peese** of lemon cake. _____

Double trouble

↻ Recap

How do I know when to double letters?

You need to learn when to use double letters and when not to.

double **r**   double **s**   no double **f**   double **s**

emba**rr**a**ss**   BUT   profe**ss**ion

📄 Revise

Let's look at words with one pair of double letters.

double **r**   double **f**   double **b**

co**rr**espond   di**ff**iculty   bu**bb**le

Learn these words in groups.
It helps you to remember them.

Now let's look at words with two pairs of double letters.

double **g**   double **s**   double **d**   double **s**

a**gg**re**ss**ive   a**dd**re**ss**

Some words even have three pairs of double letters.

double **s** double **s** double **p**

Mi**ss**i**ss**i**pp**i

💡 Tips

In words of more than one syllable, a double **consonant** usually shows that the **vowel** before it has a short vowel sound. For example: li**tt**le, mi**ss**pell.

✔ Skills Check

1. **Put a tick in the correct column to show the number of pairs of double letters.**

Word	1 pair	2 pairs
guarantee		
accidentally		
pressurised		
immediate		

2. **Write the correct spelling for each definition.**
   Each word has at least one pair of double letters.

   a. To really like something

   _____

   b. A chance to do something

   _____

   c. A decision-making group

   _____

   d. To disrupt a conversation

   _____

Some of the above words can be found in the Year 5–6 word list at the back of the book. Ask an adult to read out some words with double letters and see if you can spell them.

Tricky words

What is a tricky word?

↺ Recap

A tricky word may have:
- several **syllables**
- an unusual spelling pattern.

📋 Revise

You may need to split longer words into parts or syllables to make them easier to spell.

house: 1 syllable   **prejudice**: 3 syllables

Soft **g** sound: Is it **g** or **j**?   Soft **s** sound: **ice** or **iss**?

- Break the word into syllables (parts).
- Say each part of the word slowly and clearly.
- Then work out how to spell each syllable.

Some words have sounds which could be made in different ways.

Ask: **a** or **e**?   **sion** or **tion**?

expl**a**na**tion** = 4 syllables

Try each way. Which looks best?
Say the word clearly and you can hear the **a**.

KEY WORD

syllable

Breaking a word into syllables and then working out how to spell each part makes it easier. Try the different ways of making tricky sounds. Which looks best?

✔ Skills Check

1. Shade each syllable in a different colour. Describe the tricky bits in each word.

   a. immediately _____

   b. necessary _____

2. Circle the correct spelling of each word.

   a. government   goverment   guverment   governmeant

   b. marvelous   marvelus   marvellus   marvellous

   c. wrecognise   reckognise   recognise   reconise

Look at a word you misspell. Write the word correctly. Highlight the tricky bit and memorise the correct spelling.

Homophones

What is a homophone?

↺ Recap

A **homophone** is a pair of words which sound the same but are spelled differently.

📋 Revise

There are lots of homophones. Here are a few examples.

profit ⟶ a financial gain

prophet ⟶ someone who foretells the future

principle ⟶ a belief

principal ⟶ the leader

I awoke early this **morning**.

↗ the start of the day

They are in **mourning** following the king's death.

↗ in sorrow (following a death)

license ⟶ (verb) to allow

They were **licensed** to fish on this part of the river.

licence ⟶ (noun) a permit which allows you to do something

My television **licence** has expired.

stationary ⟶ not moving

The car was **stationary**.   (Think: There is **ar** in c**ar** and station**ar**y!)

stationery ⟶ office paper/envelopes and materials

I ordered some more **stationery** for the office.

(Think: stationery includes paper. There is **er** in pap**er** and station**er**y!)

💡 Tips

Can you find an easy way to remember what a pair of homophones mean?

For example: **here**   or   **hear**   ? Hear has **ear** hiding in it!

✔ Skills Check

1. **Underline the correct homophone for each sentence.**

   a. We **heard** / **herd** the firework display in the park.

   b. I wondered **whose** / **who's** car was parked outside my house.

   c. The burglar tried to **steel** / **steal** the television, but it was very heavy.

   d. The porridge was **two** / **to** / **too** hot!

2. **Write a sentence for each homophone.**

   a. passed _____

   past _____

   b. guessed _____

   guest _____

3. **Explain the meaning of each homophone.**

   a. aloud _____

   allowed _____

   b. farther _____

   father _____

   c. waste _____

   waist _____

4. **Write the other homophone for these words.**

   a. great _____

   b. dissent _____

   c. cereal _____

   d. bridal _____

KEY WORD

homophones

Identifying main ideas

What does identifying main ideas mean?

When you identify something, you find it in a passage. To find the main ideas, decide what a passage is about overall.

↻ Recap

The main ideas are the important things that the author wants the reader to know.

Often there will only be one main idea in a passage but there may be more than one paragraph.

📄 Revise

Don't worry about each individual idea. Look for something that links them all.

In the passage below there is one main idea.

> The house at the end of our street is very spooky.
> It is painted black and has tall, thin chimneys.
> All of the windows are dark and no one ever seems to go in or out.

Each of the sentences is about something different but they are all about the spooky house at the end of the street, so this is the main idea.

💡 Tips

- Try reading the text and then thinking of a **heading** that fits it overall.
- There are sentences on the colour of the house, what it looks like and who goes there. None of these is the main idea.
- Each sentence is about what makes the house **spooky**. So the title would be '**The Spooky House**'.

Highlight the words in each sentence that show what the sentence is about. Then find a link between them.

✔ Skills Check

1. Read this passage and identify the main idea.

> People have always been fascinated by the moon. Is it made of cheese? What is on the other side of it? Can human beings live there? Modern science has answered many of these questions and we now know that there is much more to learn about the moon than we already know.

The main idea is:

_____

_____

Identifying key details

↻ Recap

What does identifying key details involve?

- Identify means find.
- The main ideas are the important things that the author wants the reader to know.
- The key details are what the author writes about the main ideas.

📄 Revise

Start by identifying the main idea or ideas.

> The city of Hull sits proudly on the north bank of the River Humber. At one time it was the biggest fishing port in the country but now its fishing fleet has disappeared. Nowadays it is a modern city with fast motorway access and direct ferry links to Europe.

← Main idea: how Hull has changed.

Next highlight the points that tell us more about the main idea.

> The city of Hull sits proudly on the north bank of the River Humber. At one time **it was the biggest fishing port in the country** but now **its fishing fleet has disappeared**. Nowadays it is a **modern city** with **fast motorway access** and **direct ferry links to Europe**.

← Each point tells us something different.

Now, use your highlighted points to give three ways that Hull has changed.

1 It is a modern city.
2 It has fast motorway access.
3 It has direct ferry links to Europe.

✔ Skills Check

1. Read the passage below.

> Last summer we had our best holiday ever. We went to Menorca and spent a week splashing about in the pool and on the beach. We laughed all day and never had to worry about going to bed late or getting up early. I made lots of new friends.

a. What is the main idea? _____

b. Give two key details from the text to support this:

1. _____

2. _____

Summarising main ideas

↻ Recap

Summarise means sum up. When you summarise, you say briefly what the passage is about.

A summary might be one word or a complete sentence. You need to find ideas from the whole text.

📋 Revise

You have to read the whole passage before you can summarise.
In the passage below, there are different ideas for each paragraph.

My sister Carly is very kind. She has a mischievous twinkle in her eyes. She is very popular and makes every day feel like a party.

← Main idea: my sister Carly

My other sister, Caroline, is very different. She is a very private person who prefers her own company. She has a good sense of humour but rarely uses it outside of the house.

← Main idea: my sister Caroline

There are sentences about two sisters. The link between the two ideas is the difference. Put this together to summarise the main ideas of the paragraphs: the difference between the sisters.

When there is a lot of information in a passage, you might have to write more than one sentence as a summary.

The main idea in the following passage is healthy eating. The reasons that support healthy eating have been highlighted in **blue** and the reasons against it are in **orange**.

Healthy eating

Everybody loves food. Children love fast food. Burgers, chips and nuggets all taste great. There are lots of takeaway shops, meaning that fast food is easy to buy. It isn't always good for you though. Lots of fast food contains large amounts of salt and fat. Salads are really healthy but some people think that they are boring. Healthy eating gives us energy and makes us grow strong. However, if you're busy, a takeaway once in a while won't do you too much harm.

Highlight the key details and then write them in a table.

Reasons against healthy eating	Reasons that support healthy eating
Burgers, chips and nuggets all taste great.	Fast food isn't always good for you.
Fast food is easy to buy.	Fast food contains large amounts of salt and fat.
Some people think that salads are boring.	Salads are really healthy.
A takeaway once in a while won't do you too much harm.	Healthy eating gives us energy and makes us grow strong.

Use the table to help you write a summary. Concentrate on the main points.

Fast food is easy to get and it tastes great. It isn't always good for you because of what it contains. It's important to eat healthy foods like salads but a takeaway occasionally won't harm you too much.

✔ Skills check

1. **Read the passage below. Fill in the main ideas for each paragraph.**

a.

On our street there are three takeaway shops. There is an Indian, a Chinese and an Italian pizza place.

⟵ Main idea
_____

b.

We have a different meal every Saturday night. My favourites are lamb rogan josh, chicken chop suey and garlic bread.

⟵ Main idea
_____

2. **Sometimes, the summary is in the form of a heading or subheading. What do you think the best heading for the passage would be? Tick one.**

Favourite food ☐   Saturday night takeaway ☐

Our street ☐   Fast food ☐

3. **Read the passage below. Highlight the key details, then write a summary.**

My mother paints pictures. She is really good at landscapes. She's done great pictures of the sea, mountains and lakes. Her portraits aren't as good but she is working on them.

_____

Predicting what might happen

What does predicting mean?

↺ Recap

When you predict you say what you think is likely to happen. Usually you have to give reasons for your ideas. These come from clues that are written in the text.

I see. You have to read the story and say what you think will happen next. This is like being a detective.

📄 Revise

Look at this sentence:

The **fire alarm** sounded.

To predict what would happen next, you have to look for the clues in the text. In this case the clue has been highlighted. The fire alarm has gone off so what happens next must follow on from that. It has to be realistic, possible and likely. So it is no use predicting that an elephant will arrive, suck up water from a pond and then blow it down its trunk to put the fire out!

The following passage ends with the same sentence. Look at the highlighted clues to help predict what might happen next.

Mia had left the classroom to go to the toilet. On the way back she could smell **burning**. It was coming from a **store room**. Mia pushed the door open carefully and saw the **flames**. She **shut the door** quickly. **Her teacher and her classmates were in the next room**. She had to get them out. She punched the red button on the wall. The **fire alarm** sounded.

What happens next? It would be realistic, possible and likely that the teacher and the pupils would all leave the building. By closing the door, Mia has made sure that fire won't spread quickly, so that should enable the fire brigade to arrive in time to put the fire out.

Read this passage. The last sentence is different.

Mia had left the classroom to go to the toilet. On the way back she could smell burning. It was coming from a store room. Mia pushed the door open carefully and saw the flames. She shut the door quickly. Her teacher and her classmates were in the next room. She had to get them out. She punched the red button on the wall. Nothing happened!

How does changing the ending change your prediction?

You can't keep the same prediction anymore because the clue in the last sentence tells you that something else will have to happen if everyone is to be saved. This gives you a much wider choice of possible predictions.

✔ Skills Check

1. Read this passage. Highlight the important clues.

Mia hit the button again. Still nothing! She knew she mustn't panic. She ran down the corridor to her classroom and raced inside.

"The store room is on fire!" she shouted.

Mrs Milner took control. She told the pupils to leave everything on their desks and to go out of the building as quickly as possible. She made sure everyone had left the classroom and followed them. As she went outside she pressed the red button by the door. The fire alarm sounded.

Only highlight the points that give clues about what might happen. Use them to make your prediction.

a. What is likely to happen next?

_____

_____

b. Explain why you think this is likely.

_____

_____

2. Read this passage.

Once the fire was out, the chief fire officer wanted to talk to Mia. She did not know why. He interviewed her in the head teacher's office. When he spoke his tone was very serious.

a. Give two predictions about what the chief fire officer might have wanted to talk to Mia about. You should use information from the whole story in your answer.

1. _____

2. _____

b. Use evidence from the text to explain your answers.

_____

_____

_____

Themes and conventions

What are themes and conventions?

↺ Recap

- Themes are ideas that go throughout the text.
- Conventions are the things that help you know what type of writing it is.

This table shows you some of the themes and conventions.

Type of writing	Possible themes	Convention of this type of writing
Poetry	love, war	verses, rhyme, rhythm, figurative language
Drama	relationships	speech without inverted commas, stage directions
Fiction	myths and legends, adventure, love, war, good and evil, loss, fear, danger, rich and poor, strong and weak, wisdom and foolishness	heroes and heroines, villains, frightening situations, cliff-hangers, 'good' winning, using stock locations and characters – dark woods and wicked witches
Non-fiction	history, geography, celebrities, sport, gossip, cars and lots of others	textbooks, magazines/newspapers, brochures: headings, subheadings, facts, pictures, columns, bullet points, numbers and dates

You need to be able to identify themes and conventions, and comment on them.

So I need to be able to say how a text is written.

📄 Revise

In the passage below, the clues to the **theme** have been highlighted.

> Only another minute left! Karine's hands fumbled with the fuse. If she cut the wrong wire, it would be the end for all of them. Which wire? Karine didn't know! Thirty seconds. She had to choose one. A 50–50 chance. Red. No green! Fifteen seconds! This was it. Karine would have to cut one now. She held the green wire and hoped!

All of the highlighted words are typical of ones you would find in action stories. The time countdown increases the excitement. These are **conventions**. This is different to the main idea because in this paragraph the main idea would be about defusing the device, which is part of the themes of spy or war novels.

The theme of the passage below is danger. The clues that identify this have been highlighted.

> **Stranded** high on the ledge, Elliott knew **he was in trouble**. His **leg was broken** and his **radio had been lost** in the fall. Above him, a layer of ice **threatened** to drop down at any moment. He could feel himself **slipping slowly towards the edge**. It was only **a matter of time**.

To comment on the theme, explain what it is.

> **For example:** It follows the tradition that the hero faces immediate death. Everything suggests that the hero cannot survive. The ending makes us think that he will slide over the edge to his doom.

To comment on the conventions, show how they help the reader understand the theme.

> **For example:** The text is an adventure story. It has many of the usual elements including a dangerous location, an injured hero and seemingly inevitable destruction.

✔ Skills Check

To help identify the theme, highlight the ideas that go throughout the passage.

1. What is the main theme in this passage?

> Dorca the dragon flew across the night sky. Her quest was to find the secret of eternal dragon life. She knew that the knights of Nemore would try to stop her but she had dragon magic on her side!

The main theme is _____.

2. a. Find and copy a phrase that shows that the above passage has this theme.

_____

_____

b. Explain how your phrase or sentence fits this theme.

_____

c. Give two ways that the extract uses the conventions of your chosen theme.

1. _____

2. _____

Explaining and justifying inferences

↻ Recap

- Inferences are assumptions that you make from clues in the text, like how a character is feeling or why something happens. They are the bits the writer doesn't actually tell you but that you can work out for yourself.
- Explain means say what you think.
- Justify means give reasons for what you think, using parts of the text to prove your points.

What does explaining and justifying inferences mean?

OK, so it means read between the lines, work out what is happening and show us why you think that.

📋 Revise

Explaining inferences

Some parts of the text below have been highlighted. These are the clues.

Charlie was **really fed up. His day had already been bad** and he had a feeling that **it was about to get a whole lot worse**.

→ The author has told you how Charlie is feeling and why. The only thing to infer is what has made his day bad but there aren't any clues to help you.

Charlie came in from school. He **threw his bag** into the corner, **sighed loudly** and **kicked the bin**.

→ The author has not told you how Charlie is feeling or why. You have to infer that he is unhappy. The clues are in his actions. We still don't know why he is unhappy though.

1. What do the clues in the second part of the text tell us?
Charlie is behaving badly.

2. What do the clues not tell us? Why Charlie is behaving badly.

3. What inference can we make? Charlie is unhappy.

Ask yourself: 'What has happened and how do we know?'

Justifying inferences

You need to give reasons for your thoughts. To do this you need proof. This comes from the clues. In the second part of the text, the three clues that you can use as evidence have been highlighted.

What has happened to make Charlie unhappy?

The author has not told us but we can **infer** from the first sentence that something has happened at school to make Charlie unhappy.

You could try to make inferences about this but it is much more difficult as there isn't any evidence.

Writing answers

Write down an inference that you can make from the passage.

Charlie is unhappy because of something that has happened at school.

Explain an inference that you can make from the passage.

When Charlie throws his bag into the corner, sighs loudly and kicks the bin, he is showing he is unhappy. He has just come in from school so it is likely that something has happened there to upset him.

Find and copy two phrases from the text to support your inference.

1. threw his bag
2. sighed loudly

Remember to prove what you think.

✔ Skills Check

1. Read the following passage and answer the questions.

Ella went as slowly as she could into the hall. She wished she had been ill that morning. The maths test was about to take place. The test papers were lying menacingly on the desks. As Ella sat down, she could feel her heartbeat increasing. She wished she had practised more. She took a deep breath and turned the paper over.

a. How do you think Ella is feeling at the start of the passage?

_____

b. Find and copy a phrase that supports your thoughts.

_____

c. Do you think Ella will do well in the test?

_____

d. Use evidence from the text to support your thoughts.

_____

Words in context

What are words in context?

↻ Recap

Words in context means how words are used in the passage.

目 Revise

Read the following information from a text about eagles.

> There can be few more exciting sights than that of an eagle plummeting towards the earth in pursuit of its prey.

You may not know what 'plummeting' means so read the whole sentence again. The eagle is moving towards the earth. Therefore, 'plummeting' has to be telling us *how* the eagle was moving. It is in 'pursuit of its prey' so it has to be moving quickly.

What does the word 'plummeting' mean in this sentence? Tick one.

descending ☐

diving ☐

sliding ☐

tumbling ☐

In this case, you should tick diving as it fits with 'in pursuit of its prey'.

The other answers all link with 'towards the earth' but they do not give the feeling of speed.

If you don't know what a word means, try to work out what the whole sentence means and see if that gives you any clues.

✔ Skills check

Read the following passage and answer the question.

> The cliff was uneven. Slowly, bit by bit, hand over hand, I clambered up it.

1. *I clambered up it.*

Which of these words has a similar meaning to 'clambered' in this sentence? Tick one.

walked ☐   climbed ☐

raced ☐   looked ☐

Exploring words in context

What does exploring words in context mean?

↻ Recap

Explore means to go into the meaning of the words.

Now you have to look at how the words are being used as well.

📄 Revise

To explore, you have to look at a range of possible meanings of a word or phrase.
You may need to read the whole sentence or paragraph again, then work out what the word means.

Read this sentence.

> The price of bread has rocketed in the last five years.

What does 'rocketed' mean in this sentence?

The clue is in the word. What do rockets do? They move quickly; they soar upwards; they go sky-high.

To explore the use of 'rocketed', you would need to explain *how* and *why* it is being used in the sentence. **In this context rocketed has been used to show how quickly the price of bread has gone upwards because it reflects the speed of the rise.**

Tips

✔ Skills Check

If you're not sure what a word means in a sentence, read the sentences on each side of it.

Read this passage.

> I do not like rice pudding. There are few foods that I detest more. I avoid it if possible.

1. *There are few foods that I detest more.*
   Which of these words means the same as 'detest' in this sentence? Tick one.

   hate ☐   want ☐

   love ☐   need ☐

   Try all of the words and see which one makes most sense when you read the complete sentence.

2. **Explain why you have chosen your answer.**

   _____

   _____

Enhancing meaning: figurative language

What is figurative language?

↻ Recap

Figurative language is **imagery** used by writers to create word pictures that help the readers *see* what is happening and enhances the meaning.

- Examples of this include **analogy**, **metaphors**, **similes**, **personification**, **assonance** and **alliteration**.
- You need to write about the effect of the figurative language.

📄 Revise

Some of the figurative language has been identified in this passage.

> Waves **lapped** in **mournful murmurs** against the wreck of the *Free Choice*. The ship lay, **a broken-backed corpse**, across the reef. Stranded in shallow seas, the vessel **groaned like a ghost** as each wave hit it.

Figurative language	Type of language	Explain the effect
lapped	**personification** – because the waves don't really have tongues	Lapped means 'licked'. The description of the action of the waves makes us think about the movement.
mournful murmurs	**alliteration** – it creates an effect by repeating consonant sounds, in this case the 'm'	Alliteration helps the reader hear the noise.
a broken-backed corpse	**metaphor** – it **compares more strongly**, usually using **'is'** or **'was'**. The metaphor also uses **personification** to help the reader picture the ship	This metaphor helps us imagine how damaged the ship looks.
groaned like a ghost	**simile** – it **compares** by using **'like'** or **'as'** **alliteration**	The simile compares the groaning to a ghost, which makes the description more vivid.

It's important to know the name of the figurative language but it's even more important to say what its effect has been.

✔ Skills check

1. Read the following passage.

> My brother had broken my favourite toy.
> I roared like a monster in anguished anger.
> Tears burned my eyes. My mother, hearing
> my cries, held me like a nurse until I stopped
> sobbing. My father brought the remedy –
> superglue.

a. *I roared like a monster…*
The sentence above contains: Tick **one**.

a metaphor ☐

alliteration ☐

personification ☐

a simile ☐

KEY WORDS

figurative language
imagery
analogy
metaphors
similes
personification
assonance
alliteration

b. In the table below, highlighted in bold, are examples of figurative language from the passage. In each case, state what type of figurative language is used and explain its effect.

Language used	Type of language and effect
*Tears **burned** my eyes*	
*held me **like a nurse***	
*My father brought the **remedy – superglue**.*	

How writers use language

⟳ Recap

Writers use language to have an effect on the reader through:
- vocabulary used
- use of different sentence types and links between them
- different types of text (fairy stories, newspapers, magazines, letters).

You have to write about the effect each has on the reader.

How do writers use language?

Always explain why and how the language that is used affects the reader.

📄 Revise

Words

Different words show different shades of meaning. Some words like 'nice' or 'good' are very vague and give the reader little idea of what they mean. 'It was a scary film' could cover anything from mildly creepy to bloodcurdlingly terrifying. Choosing words carefully is important to ensure that readers know exactly what writers are trying to say.

Sentences

Different forms of sentence create a response in the reader.
- **Everybody loves ice cream.** This is a **statement**. It seems to be a fact but is it? Actually, it's not a fact but it is presented as one so the effect is that the reader believes it is true.
- **Who can argue that Britain is the best country in the world?** This is a **rhetorical question**. It tries to make the reader agree with it by suggesting that no one could argue against it.
- **Finish ahead. Foot flat on the floor! Maximum speed. Go! Go! Go!** These **short sentences** have the effect of making the action seem fast, almost like a series of photographs.

Text

Texts are written for different purposes. You need to be able to identify the purpose and show how the writing fits it. This one is written to inform.

This includes you!

Reasons to exercise are highlighted: 'makes us healthier', 'live longer', 'feel better', 'reduce weight'

Everyone needs to exercise. Exercise makes **us healthier** and can help **us live longer**. It can help us feel **better** and **reduce weight. You** could walk, jog, run or swim. These are cheap and they will all make a huge difference to **your** health. For further information, visit www.healthierlife.com

Use of 'us' and 'you' makes it seem like the writer is talking to the reader

Gives information and choices: walk, jog, run or swim

More information

Place to find more information

✔ Skills Check

1. The following questions are about the Exercise extract.

a. Give four reasons why we should exercise.

_____

_____

b. Name two forms of exercise that are recommended.

_____

_____

c. *These are* **cheap** *and they will all make a* **huge** *difference to your health.*
What is the effect on the reader of the words in bold?

_____

_____

2. Read this passage.

> Run! Run like your life depends on it – because it does.
> Run! Don't look back! Run! You'll know when to stop.

Language used	Effect of language
a. *Run!*	What is the effect on the reader of repeating *Run!* _____ _____
b. *because it does*	What is the effect on the reader of this phrase? _____ _____
c. *Don't look back!*	How does this sentence increase the tension or excitement in the passage? _____ _____

💡 Tips

For this question, you have to show which words and writing techniques have been used and what the effect is on the reader.

Features of text

What are features of texts?

↺ Recap

- Language features – the way the words are used.
- Structural features – the way the text is organised.
- Presentational features – the way the text looks.

📄 Revise

Here are some examples of different features.

Language features	Structural features	Presentational features
• Figurative language • Short/long sentences • Variety/repetition of words • Specific choice of words • Rhetorical questions	• Chapters • Table of contents • Headings and subheadings • Paragraphs or verses	• Pictures and captions • Diagrams • Columns and charts • Text boxes • Fonts and colour

In the following passage, a number of language, structural and presentational features have been identified for you.

heading and bold text

metaphor

The Damage is Done

A **whirlpool of emotion** splashed tears into Sara's eyes. Why did it have to happen *now*? Things had been going **like clockwork**. Lewis wasn't meant to get that text. It was stupid. **Stupid, stupid, stupid!**

italics

simile

repetition

✔ Skills check

Why oh why?

Sara stared at her phone. Why had she hit the *send* button?

1. Find and copy examples of language, structural and presentational features from the above text.

Feature	Feature name	Example
Language		
Structural		
Presentational		

Text features contributing to meaning

How do text features contribute to meaning?

↻ Recap

Text features are the language, structural and presentational features of texts.

You need to explain how they help the reader understand the meaning of the text.

📄 Revise

This is the continuation of **The Damage is Done**. Some features are highlighted.

subheading and bold text →

short sentence →

repetition

simile →

> **Be careful what you wish for**
>
> It was a joke!
>
> She never meant to send it. She never meant Lewis to get it. She never wanted anyone to get hurt. Her phone shook like an earthquake in her hand. It was a reply from Lewis.

You need to be able to explain how each feature works in the passage as a whole.

Feature	What it does	Explanation
Subheading	Makes it easy to read	Breaks up the text and gives a summary of the main idea of the paragraph
Bold text	Draws attention to important text	Makes it stand out
Short sentence	Increases pace of text	It tells us more about what Sara was doing in a short space of time
Repetition	Reinforces point	Emphasises how little Sara had wanted the events to take place
Simile	Helps the reader imagine the scene	An earthquake is a huge disaster. That is what Sara is expecting when the text comes in

✔ Skills Check

Read both parts of 'The Damage is Done'.

1. How does writing in the third person help the reader?

_____

2. *It was a reply from Lewis.*

How does the writer build up tension in this sentence?

_____

Retrieving and recording information

What does retrieving and recording information mean?

↺ Recap

- Retrieve means find.
- Record means write down.

📄 Revise

Read the following passage. Key pieces of information have been highlighted.

The *Titanic*

The *Titanic* has captured the imagination of the public more than any other ship in history. Perhaps it is because it was described as 'unsinkable' by its designer. Perhaps it is because it sank on its first voyage. Perhaps it is because there is so much mystery surrounding its loss. Whatever the reason, there has been continuous interest in the *Titanic* for over a hundred years.

Look for the first key words in the question: why, what, who, where, when or how.

Example questions

What are the reasons people have been interested in the *Titanic*?

Is the same as:

Why have people been interested in the *Titanic*?

Some questions will ask you to join information together. For example, in this case, you might be asked to **draw lines to link the *Titanic* to why people might be interested in it.**

Look for the **other key words in the question**: these tell you what to retrieve from the text. In this question they are: 'interested in the *Titanic*'.

Scan the text above and you'll find three reasons.

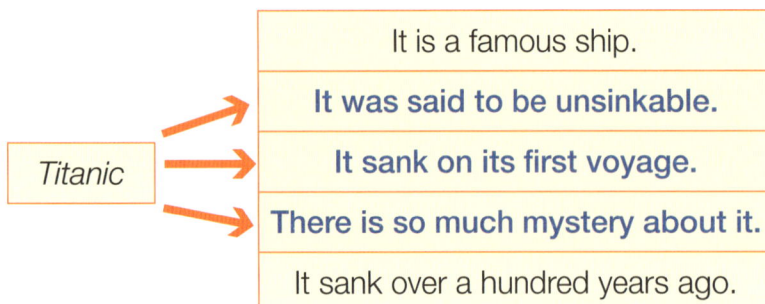

It is a famous ship.
It was said to be unsinkable.
It sank on its first voyage.
There is so much mystery about it.
It sank over a hundred years ago.

Titanic →

Tips 💡

Look closely to find the answers.
- **Why** = find a reason
- **Who** = find a name
- **Where** = find a place
- **When** = find a time
- **How** = find an explanation
- **What** can be any of the above.

✔ Skills Check

1. Read 'The *Titanic*' again.

a. Who is interested in the *Titanic*?

_____

b. How long is it since the *Titanic* sank?

_____

c. Why would people be surprised that it sank?

_____

d. Interest in the *Titanic* has been: Tick **one**.

increasing. ❏

reducing. ❏

continuous. ❏

overwhelming. ❏

> Remember to look for the key words. They tell you what to retrieve.

2. Read the continuation of 'The *Titanic*'.

Nowadays the *Titanic* lies at the bottom of the Atlantic Ocean. It is still recognisable as the wonderful ship it once was, even though it is encrusted with barnacles and sea life. Instead of the rich and the famous, it is now home to a whole host of different sea creatures. There are memorials to the *Titanic* in Belfast, Liverpool, Southampton, Washington DC and New York. Although there is still huge interest in it, it will probably never be brought to the surface.

Draw lines to link the *Titanic* with information about it nowadays.

It is at the bottom of the Pacific Ocean.
It looks just like it did before it sank.
Sea creatures now live in it.
There are memorials in five cities.
It will be raised soon.

Titanic

Making comparisons

What does making comparisons mean?

Comparisons show us what is similar or different in a text.

📋 Revise

Read the following passage.

My favourite birds

My hobby is bird watching, or ornithology, to give it its proper name. Why have two names? Well, it's a bit like the birds themselves. They all have common names but they also all have proper names. Did you know that sparrows are also called Passeridae? No? Neither did I until I started watching them.

I don't really like little birds. My favourites are the hunting birds. I love owls. They are so graceful. They fly in silence, seemingly without effort. Some people don't like them at all. They see them as cruel hunters. They prefer birds that don't even fly, like penguins, but they hunt too. People like the awkward way they walk. What's the point in a bird that can't fly? It's like a fish that's scared of water.

Give one reason why people might like owls and one why they might not.

Like owls: They are graceful or they fly in silence, seemingly without effort.

Dislike owls: They are seen as cruel hunters.

Read the continuation of the passage.

Owls glide through the air in pursuit of their prey. Penguins can't fly but they do the same underwater. An owl in flight is a graceful sight. Penguins are equally graceful in the water. Owls hop while penguins walk. Neither bird is particularly comfortable on the ground but of the two, penguins are more suited to being on land. Owls can be found in the wild all over the world but penguins only live in the southern hemisphere.

To compare, you have to show the similarities and the differences.

Comparison	Owls	Penguins
Similarities – movement	Graceful in flight	Graceful in water
Differences – movement	Hop	Walk
Differences – where they live	All over the world	In the southern hemisphere

✔ Skills check

1. Read the continuation of the 'My favourite birds' passage again.

   a. Give one thing that is different between owls and penguins.

   _____

   b. Give one thing that is similar between owls and penguins.

   _____

2. Read the following passage.

> Some animals are more popular than others. For instance, cats are almost universally liked while snakes are hated the world over. I am perfectly happy to have my cat slide onto my knee and settle down to watch television with me. I can't say the same of a snake. Why? Cats are much more familiar to us so their danger seems less. Cats do not hunt large prey, like us. Snakes? Well, what do we know? They hiss and spit but so do cats when they are annoyed. Snakes hunt small animals. So do cats.

> *Don't get yourself tied up in knots. To compare just show similarities and differences.*

   a. Compare the author's attitude to cats and snakes.

   _____

   _____

   _____

   b. What is similar between cats and snakes?

   _____

   _____

   c. What differences are there between cats and snakes?

   _____

   _____

Find the key words in the question and look for them in the text.

Fact and opinion

↻ Recap

What is fact and opinion?

- A **fact** is true and can be proved.
- An **opinion** is what someone thinks or believes. You need to be able to tell the difference between facts and opinions.

To tell if something is a fact, ask: 'Can it be proved?'

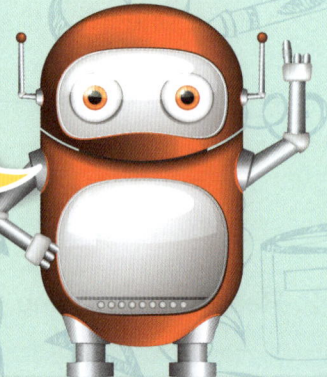

Revise

In the passage below, there is one **fact** and one **opinion**.

> Jessica Ennis-Hill won a gold medal in the heptathlon at the London Olympics in 2012. She is the most talented heptathlete Britain will ever have.

- Jessica Ennis-Hill did win the Olympic gold medal. **Can this be proved? Yes.** There is video evidence to prove it. She is the most talented...will ever have. **Can this be proved? No.** Nobody knows what will happen in the future.

- The text makes it seems as if Jessica is the most talented heptathelete. How? It starts with a fact that proves that she is a top-class athlete and therefore implies she is the most talented. We might believe the second sentence is a fact because we know that the first one is but there is no way we can predict the future and what new heptathletes Britain will have in the future.

Read the following passage.

> The tallest mountain in Wales is Mount Snowdon. It is known in Welsh as Yr Wyddfa.

How many **facts** are in the passage? **Two.**
Both sentences contain **facts** that can be proved.

> At 1085 metres high, Yr Wyddfa is the highest mountain in Wales. You can walk to the top or, for the less energetic, there is the Snowdon Mountain Railway.

Tips

Watch out for **'foggy phrases'**. They're hard to see through and it's easy to get lost in them! 'Everybody knows' and 'There can be no doubt' are foggy phrases. They make things seem like **facts** when really they are **opinions**.

✔ Skills Check

Read the following passage.

> I think that the Mona Lisa is a strange picture. Painted by Leonardo da Vinci and also called La Gioconda, it is kept in the Louvre in Paris. It is only 77cm by 53cm and is hung in a dark room to avoid the light damaging it. When I went, the room was packed with people trying to see the picture. It was not awe-inspiring. The statue of Venus de Milo is much more impressive. It's not worth queuing to see the Mona Lisa. You'd be better off spending your time in the Egyptian section. The sphinx in there is really impressive.

1. Put a tick in the correct box to show whether each of the following statements are fact or opinion.

	Fact	Opinion
The Mona Lisa is 77cm by 53cm.		
The Mona Lisa is a strange picture.		
The Mona Lisa is not awe-inspiring.		
The Mona Lisa is in a dark room.		

> Sentences that include 'I think' or 'I believe' are opinions. Don't be fooled. Remember, facts can be proved.

2. Find and copy three facts from the passage that are not included in the table above.

   1. _____

   2. _____

   3. _____

3. Find and copy three opinions that are not included in the table above.

   1. _____

   2. _____

   3. _____

Maths
SATs Made
Simple
Ages 10–11

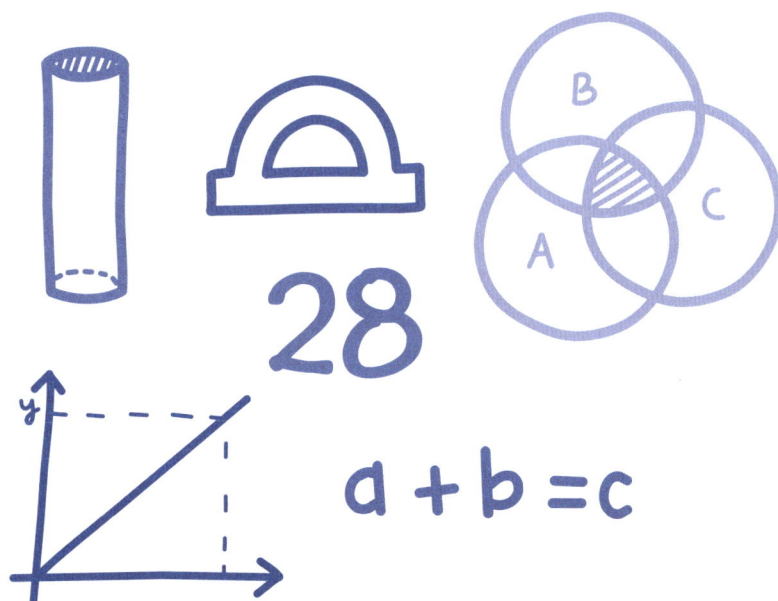

28

$a + b = c$

The number system

↻ Recap

In the past, some people used the Roman system when writing numbers. The Romans used letters to represent amounts.

Nowadays we use ten digits:

0 1 2 3 4 5 6 7 8 9

All of the maths we do only uses these ten digits.
We can do a lot with only ten digits because of **place value**.

📄 Revise

Our number system is called **base 10** because it arranges digits in columns that increase in powers of 10.

1,000,000s	100,000s	10,000s	1000s	100s	10s	1s
5	6	4	0	3	4	2

Notice how each digit represents a different amount depending on its place value.

💡 Tips

- Make sure you can read large numbers. It isn't so hard if you take your time. Look at this number:

2450398

- We can separate the millions and thousands using gaps or commas.

2,450,398 or 2 450 398

- If you are still unsure, write in the place value above each digit.

1,000,000s	100,000s	10,000s	1000s	100s	10s	1s
2	4	5	0	3	9	8

Say it out aloud: two million, four hundred and fifty thousand, three hundred and ninety-eight.

There's certainly a place for gaps or commas!

💬 Talk maths

DID YOU KNOW?

Romanian is the closest living language to Latin, the language of the Romans.

Roman numerals

Work with a partner. Challenge each other to say any Roman numeral up to 1000. The table below gives you everything you need to know.

Number	1	2	3	4	5	6	7	8	9	10
Roman numeral	I	II	III	IV	V	VI	VII	VIII	IX	X
Number	50	100	500	1000						
Roman numeral	L	C	D	M						

Base 10 numbers

Starting small and getting bigger, write down ten numbers up to 10,000,000 and challenge your partner to say them correctly.

349   9235   400,004   45,202

305,621   3,452,320   90,009

3,000,003   6,426,208

726,817

It's quite easy once you get the hang of it!

✔ Check

1. Change these Roman numerals to base 10 numbers.

   a. CCCL _____   b. CXC _____   c. MMMD _____   d. MDCLXVI _____

2. Write the value of the underlined digit in each number.

   a. 32,4̱02 _____   b. 2̱30,508 _____

   c. 4̱,730,627 _____   d. 7,6̱73,205 _____

⚠ Problems

Brain-teaser Write the number that is one more than one million. _____

Brain-buster What is the biggest 7-digit number? Write it in digits then in words.

_____

_____

Numbers to 10,000,000

↻ Recap

239,718 in words is two hundred and thirty-nine thousand, seven hundred and eighteen.

100,000s	10,000s	1000s	100s	10s	1s
2	3	9	7	1	8

The **place value** of the three digit represents 30,000; the seven represents 700. What do the other digits represent?

Zeros are also important.
402,005 in words is four hundred and two thousand and five.

100,000s	10,000s	1000s	100s	10s	1s
4	0	2	0	0	5

🗒 Revise

This number is twelve million, seven hundred and sixty-four thousand, three hundred and five.

10,000,000s	1,000,000s	100,000s	10,000s	1000s	100s	10s	1s
1	2	7	6	4	3	0	5

Use commas after the millions and after the thousands column. The number above should be written as 12,764, 305.

What number does each of the digits represent?

💡 Tips

- Write the place value in columns above numbers if you're stuck.
- < means less than and > means more than.

DID YOU KNOW?

A billion is a thousand million. One billion has nine zeros.

?

Talk maths

With a partner, practise saying these numbers.

10,000	10,000,000	5,999,999 < 6,000,001
100,000	7,291,428	2,450,312 > 1,974,489
1,000,000	21,426,3007	9,999,999

DID YOU KNOW?

1000^2 (one thousand squared) = 1,000,000

✔ Check

1. Write this number in words. 845,283

   _____

2. Write this number using digits.

   six hundred and four thousand, one hundred and ninety _____

3. What does the 6 digit represent in 3,682,309? _____

4. Put these numbers in order, from smallest to largest.

   825,421   10,000,000   97,612   6,899,372   500,000

   _____   _____   _____   _____   _____

5. Insert the < or > sign between each pair to make the number statements correct.

   a. 3521 _____ 5630   **b.** 15,204 _____ 9798   **c.** 833,521 _____ 795,732

⚠ Problems

City	Rome	Paris	Madrid
Population	2,646,346	2,341,895	3,324,031

Brain-teaser Which city has the largest population? _____

Brain-buster Write the names of these cities in order, from smallest population to largest population.

_____   _____   _____

Estimation and rounding

↻ Recap

We use powers of 10 for rounding, counting and estimating.

To round a number to the nearest **power of 10** we look at it on a number line.

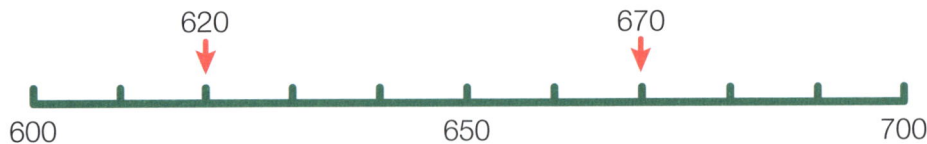

620

670

600   650   700

620 rounded down to the nearest hundred is 600

670 rounded up is 700

649 and below will round down to 600; 650 and above round up to 700.

We can count on in 100s too: 100, 200, 300, 400, 500 and so on.

And we can also use these skills to estimate answers, for example, 103 + 98 + 204 + 195 = is approximately 100 + 100 + 200 + 200 = 600

▤ Revise

We can do the same with thousands and millions.

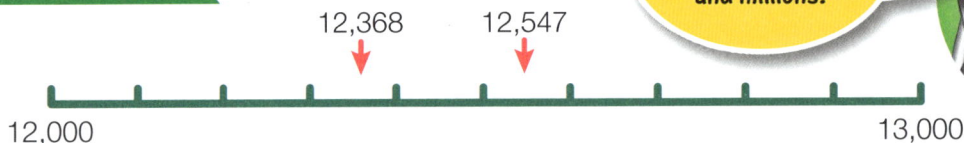

12,368   12,547

12,000   13,000

12,368 rounded down to the nearest thousand is 12,000

12,547 rounds up to 13,000

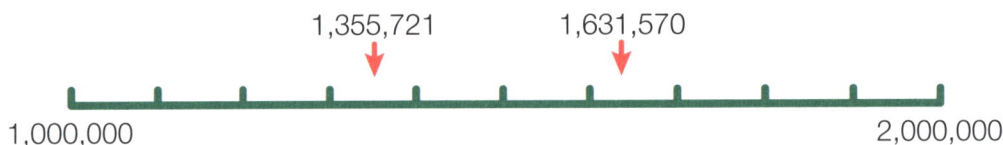

1,355,721   1,631,570

1,000,000   2,000,000

1,355,721 rounded down to the nearest million is 1,000,000

1,631,570 rounds up to 2,000,000

To estimate the answer to 45,231 + 23,876 we could say 45,000 + 24,000 = 69,000.
To estimate the answer to 7,235,421 − 5,862,403 we could say
7,000,000 − 6,000,000 = 1,000,000.

💡 Tips

Always think carefully about what you want to round to: thousands, ten thousands, millions, and so on. Then think about the part of the number line the number is on.

Talk maths

Work with a partner. Each write six different numbers between 10,000 and 10,000,000. Say aloud each other's numbers and then challenge each other to round any of the numbers to a power of 10.

What is 5,348,325 rounded to the nearest 100,000?

✔ Check

1. Round these numbers to the nearest 1000.
   a. 4567 _____
   b. 23,145 _____
   c. 45,320 _____
   d. 78,649 _____

2. Round these numbers to the nearest 100,000.
   a. 120,367 _____
   b. 450,000 _____
   c. 1,382,320 _____
   d. 7,976,311 _____

3. Round these numbers to the nearest 1,000,000.
   a. 6,435,207 _____
   b. 845,453 _____
   c. 3,500,000 _____
   d. 9,724,500 _____

4. Complete these sequences.
   a. 0; 100,000; 200,000; _____; _____; _____;
   b. 370,000; 380,000; _____; _____; _____;
   c. 7,500,000; 8,500,000; _____; _____; _____;

⚠ Problems

Brain-teaser Round each city's population to the nearest million.

City	Rome	Paris	Madrid
Population	2,646,346	2,341,895	3,324,031

Rome _____   Paris _____   Madrid _____

Brain-buster Estimate, to the nearest million, the total population of Madrid, Rome and Paris.

_____

Do you think your estimate is higher or lower than the actual total? Explain your answer.

_____

_____

97

Negative numbers

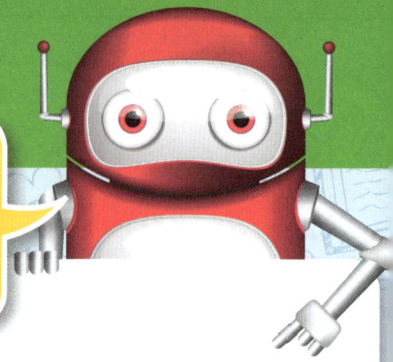

↺ Recap

When we add numbers on a number line we move to the right. When we take away numbers we move to the left.

Numbers can be negative as well as positive.

| –10 –9 –8 –7 –6 –5 –4 –3 –2 –1 0 1 2 3 4 5 6 7 8 9 10 |

Remember you can use a number line to help you. Don't forget to include zero when you are counting!

📄 Revise

Temperature is a great way to practise using positive and negative numbers.

If you start at +2 and count back 6 you stop at −4.

If you start at +15 and count back 16 you end at −1.

If you start at −8 and count on 16 you stop at +8.

If you start at −13 and count on 24 you stop at +11.

We can do simple calculations with positive and negative numbers to check the answer. For example:

$2 - 3 = -1$ **so** $-3 + 2 = -1$ $-14 + 18 = 4$ **so** $18 - 14 = 4$

You just need to remember to swap the minus to a positive and the positive to a negative.

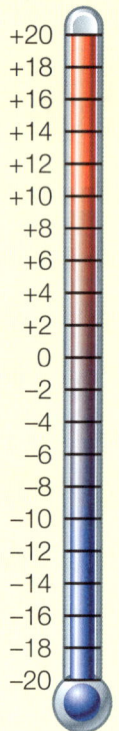

```
+20
+18
+16
+14
+12
+10
+8
+6
+4
+2
0
−2
−4
−6
−8
−10
−12
−14
−16
−18
−20
```

💡 Tips

- Can you spot the connections between positive and negative numbers? Look at the connections in the box. If you understand this, negative numbers will be easy for you!

$8 - 4 = 4$	$4 + 8 = 12$	$12 - 8 = 4$	$12 - 8 = 4$
$4 - 8 = -4$	$-4 - 8 = -12$	$8 - 12 = -4$	$4 - 12 = -8$

- Try choosing some other numbers and see if you can spot patterns.

Talk maths

7	9	12
4	16	20
−3	−12	−8
−20	−17	−5

What's minus ten plus fifteen?

Minus five.

With a partner, choose two numbers from the box and ask them to either subtract or add them together. For example, say: *What is nine minus twelve?* Now ask them to ask you some questions. Use the number line below to help you.

−20 −19 −18 −17 −16 −15 −14 −13 −12 −11 −10 −9 −8 −7 −6 −5 −4 −3 −2 −1 0 1 2 3 4 5 6 7 8 9 10 11 12 13 14 15 16 17 18 19 20

✔ Check

1. Complete these calculations.

    a. $3 - 5 =$ _____   **b.** $5 - 9 =$ _____   **c.** $-4 + 7 =$ _____   **d.** $-8 + 8 =$ _____

2. Count on from −20 to +20 in steps of 4. Write each number.

    _____ _____ _____ _____ _____ _____ _____ _____ _____ _____

3. Write the missing signs + or −.

    a. 7 _____ 7 = 0   **b.** −12 _____ 13 = 1

    c. 14 _____ 21 = −7   **d.** 2 _____ 18 = −16

4. Write the missing numbers.

    a. $-13 +$ _____ $= 1$ **b.** $14 -$ _____ $= -5$ **c.** _____ $- 15 = -8$ **d.** _____ $+ 10 = 1$

⚠ Problems

Brain-teaser One winter morning the temperature at dawn is −4 degrees Celsius (−4°C). If the temperature rises 12°C by noon, what will the temperature be then?

_____

Brain-buster The temperature in the desert 49.7°C and in the mountains is −19.7°C.

What is the difference between the two places? _____

DID YOU KNOW?

Even though deserts are hot places, they can get very cold at night.

Addition and subtraction

To add 999, just add 1000 and subtract 1.
45,362 + 999 = 46,361.

↺ Recap

You will probably know several mental methods for addition and subtraction.

You must learn your number bonds: 7 + 8 = 15   15 − 8 = 7   15 − 7 = 8
Partitioning numbers is important too: 25 + 12 = 37

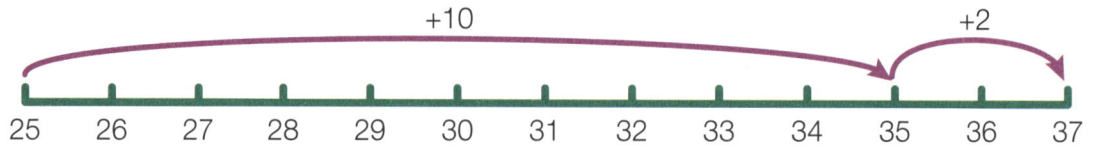

📄 Revise

We can use formal written methods for adding and subtracting larger numbers.

The first step is to neatly lay out the numbers in columns according to place value.

```
    6 6 4 5 7 2
+   1 5 3 0 5 4
─────────────────
    8 1 7 6 2 6
      ₁     ₁
```

Just like addition, we can use the place-value columns to subtract larger numbers.

```
  ²3̶ ¹³4̶ ¹⁰5̶ ¹2 4 6
−   1 6 5 3 0 4
─────────────────
    1 7 5 9 4 2
```

If in doubt, ask someone to show you.

💡 Tips

- Remember, you can check your subtractions by adding your answer to the number you took away.

```
  ¹2̶ ¹3 ³4̶ ¹³4̶ ¹3
−     6 1 7 5
─────────────────
    1 7 2 6 8    checking…
```

```
    1 7 2 6 8
+     6 1 7 5
─────────────────
    2 3 4 4 3    correct! ☺
      ₁   ₁ ₁
```

- Only use written methods that you are sure you understand. If you have a method you like, stick to it, practise it, and always check your answers!

Talk maths

Try it with three numbers, or even four!

Think of two numbers and write them down. Challenge a friend to add them using a mental or written method, and then explain their method to you. Repeat this five or six times, then do the same for subtractions.

✔ Check

1. **Add these numbers using mental methods.**

   a. 452 + 340 = _____   **b.** 5127 + 399 = _____   **c.** 425,364 + 54,005 = _____

2. **Subtract these numbers using mental methods.**

   a. 800 − 260 = _____   **b.** 146,450 − 29,000 = _____   **c.** 2,754 − 399 = _____

3. **Add these numbers using a written method.**

   a. 234,482 + 314,222   **b.** 635,231 + 327,594   **c.** 1,342,435 + 3,825,032

4. **Subtract these numbers using a written method.**

   a. 314,222 − 234,482   **b.** 962,825 − 327,594   **c.** 3,825,032 − 1,342,435

⚠ Problems

City	Rome	Paris	Madrid
Population	2,646,346	2,341,895	3,324,031

Brain-teaser How many more people live in Madrid than Paris? _____

Brain-buster Calculate the combined population of Rome, Paris and Madrid. _____

Multiplication and division facts and skills

↺ Recap

Multiplication squares show us that division is the *inverse* of multiplication.

So, we can say:
$8 \times 9 = 72$
$9 \times 8 = 72$

$72 \div 9 = 8$
$72 \div 8 = 9$

×	1	2	3	4	5	6	7	8	9	10	11	12
1	1	2	3	4	5	6	7	8	9	10	11	12
2	2	4	6	8	10	12	14	16	18	20	22	24
3	3	6	9	12	15	18	21	24	27	30	33	36
4	4	8	12	16	20	24	28	32	36	40	44	48
5	5	10	15	20	25	30	35	40	45	50	55	60
6	6	12	18	24	30	36	42	48	54	60	66	72
7	7	14	21	28	35	42	49	56	63	70	77	84
8	8	16	24	32	40	48	56	64	72	80	88	96
9	9	18	27	36	45	54	63	72	81	90	99	108
10	10	20	30	40	50	60	70	80	90	100	110	120
11	11	22	33	44	55	66	77	88	99	110	121	132
12	12	24	36	48	60	72	84	96	108	120	132	144

📋 Revise

You already know some square and cube number facts, and you can calculate others.

Five squared = $5^2 = 5 \times 5 = 25$ Five cubed = $5^3 = 5 \times 5 \times 5 = 125$

Remember the inverses: $25 \div 5 = 5$, $125 \div 5 = 25$

Also, you should now be able to multiply and divide by **powers of 10**.

Operation	Fact	Example
×10	Move one place left	$65 \times 10 = 650$
÷10	Move one place right	$65 \div 10 = 6.5$
×1000	Move three places left	$65 \times 1000 = 65,000$
÷1000	Move three places right	$65 \div 1000 = 0.065$
×1,000,000	Move six places left	$65 \times 1,000,000 = 65,000,000$
÷1,000,000	Move six places right	$65 \div 1,000,000 = 0.000065$

💡 Tips

When multiplying by larger numbers, we can separate the powers of 10, for example:

$7 \times 12,000$ is the same as $7 \times 12 \times 1000$
$= 84 \times 1000 = 84,000$

Or   for $24,000 \div 6$, just do $24 \div 6 = 4$, then times by 1000
$= 4 \times 1000 = 4000$

🗨 Talk maths

Try to out-smart an adult. Ask them to solve a calculation mentally, then give them a challenge such as to multiply a square or cube number by a power of 10. For example:

What is seven squared times a thousand?

What is three cubed times one hundred thousand?

If you are feeling brave, work out some answers in advance and then try out-smarting an adult with a mental division, for example:

What is forty-nine thousand divided by seven?

What is two thousand seven hundred divided by three?

✔ Check

1. Solve these multiplications mentally.

   a. $24 \times 200 =$ _____

   b. $62 \times 1000 =$ _____

   c. $40 \times 40 =$ _____

   d. $25 \times 2000 =$ _____

   e. $43 \times 10,000 =$ _____

   f. $100 \times 10,000 =$ _____

2. Now solve these divisions using mental methods.

   a. $6000 \div 3 =$ _____

   b. $125 \div 5 =$ _____

   c. $120,000 \div 3 =$ _____

   d. $360,000 \div 4 =$ _____

   e. $640,008 \div 8 =$ _____

   f. $125,000 \div 5 =$ _____

3. Use your knowledge of inverses to solve these.

   a. If $27,072 \div 576 = 47$, what does $576 \times 47 =$ _____

   b. If $4320 \times 723 = 3,123,360$, what does $3,123,360 \div 4320 =$ _____

⚠ Problems

Brain-teaser A football stadium holds 8000 people. How much money would be collected for a sell-out match if each ticket was £20?

_____

Brain-buster For a different football match, tickets are sold for £30, but only £90,000 is collected.

How many tickets were sold? _____

Written methods for long multiplication

↺ Recap

There are several formal written methods for multiplying numbers. You may have been taught methods a bit different from this one. You should use whichever method you are comfortable with.

		3	6	
	×	2	4	
	1	4²	4	(← × 4)
	7¹	2	0	+ (← × 20)
	8	6	4	

Answer: 864

		4	7	
	×	1	8	
	3	7⁵	6	(← × 8)
	4	7	0	+ (← × 10)
	8	4	6	

Answer: 846

Remember, the numbers are arranged in their place-value columns: hundreds, tens and ones.

📄 Revise

We can use formal written methods for all numbers, no matter how large they are.
Multiplying two numbers that are both larger than 10 is called long multiplication.
We multiply each digit on the top by each digit on the bottom, carrying forward powers of 10.

			3	2	6	
		×		4	5	
		1	6¹	3³	0	(← × 5)
	1	3¹	0²	4	0	+ (← × 40)
	1	4	6	7	0	

Answer: 14,670

			4	2	0	8	
		×			6	3	
		1	2	6	2²	4	(← × 3)
	2	5¹	2	4⁴	8	0	+ (← × 60)
	2	6	5	1	0	4	

Answer: 265,104

Remember, always put the larger number on the top.

💡 Tips

Lay out your work neatly and you'll probably get the right answer.

Watch how to do huge calculations and get them right!

		8	6	9	5	
	×		6	7		
6	0⁴	8⁶	6³	5		(← × 7)
5	2⁴	1⁵	7³	0	0	+ (← × 60)
5	8	2	5	6	5	

Answer: 582,565

Talk maths

Look at each of these long multiplications and talk it through aloud, explaining how each stage was done. Make sure you work in the correct order.

		4	8
	×	3	1
		4	8
1	4²| 4	0	+
1	4	8	8

Answer: 1488

		6	0	7
	×		2	5
	3	0	3³	5
1	2	1¹| 4	0	+
1	5	1	7	5

Answer: 15,175

Remember that zeros still have to be multiplied and recorded, and anything times zero is... zero!

✓ Check

1. Complete each of these long multiplications using a written method.

a. 62 × 14

b. 325 × 22

c. 405 × 34

d. 6338 × 52

2. Complete each of these long multiplications using a written method.

a. 425 × 21

b. 1267 × 30

c. 5326 × 15

d. 8736 × 65

⚠ Problems

Brain-teaser A head teacher estimates that every child in her school does 72 pieces of homework each year (that is around two pieces per week). If there are 347 children in the school, how many pieces of homework must be marked each year?

_____

Brain-buster A supermarket chain sells 9237 RoboDog toys in a year. They cost £79 each.

How much money does the supermarket make in total? _____

Written methods for short division

↺ Recap

Remember what divide means. It tells you how many times one number goes into another number.

There are several formal written methods for dividing numbers. You may have been taught methods a bit different to those in this book. You should use whichever method you are comfortable with – as long as you get the right answers!

		0	8	6
3	2	²5	¹8	

Answer: 258 ÷ 3 = 86

For 72 ÷ 8 = 9 we say, 72 **divided by** 8 equals 9.

📋 Revise

In short division we carry forward remainders. Sometimes there is a remainder in the answer at the end.

		0	6	3	r1
4	2	²5	¹3		

Answer: 253 ÷ 4 = 63 r1

		0	2	2	5	r2
7	1	¹5	¹7	³7		

Answer: 1577 ÷ 7 = 225 r2

You can learn about long division in the next unit.

💡 Tips

I'll keep this tip short – get it?!

- Lay out your work carefully and think about the place value of every digit. Use squared paper to help you.

		0	8	5	8	6	9	r1
3	2	²5	¹7	²6	²0	²8		

Answer: 257,608 ÷ 3 = 85,869 r1

- You can check your answer by multiplying the answer by the number you divided by, and then add the remainder. Look:

	8	5	8	6	9
×					3
2	5	7	6	0	7
	1	2	2	2	

Answer: 257,607 + 1 remainder = 257,608

💬 Talk maths

Look at this short division and explain it aloud, saying how each stage was done.

	1	4	5	2	0	0	5	r4
6	8	²7	³1	¹2	0	3	³4	

Answer: 8,712,034 ÷ 6 = 1,452,005 r4

✔ Check

1. Complete each of these short divisions.

a. 92 ÷ 4

b. 123 ÷ 5

c. 2605 ÷ 6

d. 3758 ÷ 12

2. Complete each of these short divisions using a written method.

a. 86 ÷ 7

b. 322 ÷ 5

c. 3685 ÷ 8

d. 13,588 ÷ 12

⚠ Problems

Brain-teaser A teacher shares out 93 stickers between seven children. How many stickers will each child receive, and how many will be left over? _____

Brain-buster Tickets for a pop concert cost £18 each. If the total amount taken for tickets was £22,464,

how many tickets were sold? _____

Explain how you could check your answer.

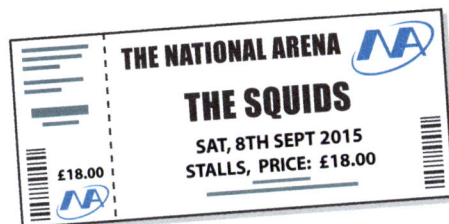

THE NATIONAL ARENA

THE SQUIDS

SAT, 8TH SEPT 2015
STALLS, PRICE: £18.00

£18.00

_____

_____

Written methods for long division

Turn back a page to see formal methods for short division.

↺ Recap

To divide something means to share it into equal amounts. Twelve divided by three equals four.

For larger numbers we sometimes need to use formal methods to help us calculate accurate answers.

In short division we carry on the remainder at each stage.

	0	4	2	6	r2
8	3	³4	²1	⁵0	

Answer: 426 r2

📄 Revise

When we are dividing larger numbers we may need to use long division. This example shows you one method.

Can you see the difference between long division and short division? With long division we are calculating the remainder at each stage, so that there is less chance of making an error.

				2	2	3	r3
		1	6	3	5	7	1
(2 × 16 =)	−			3	2		
					3	7	
(2 × 16 =)	−				3	2	
						5	1
(3 × 16 =)	−					4	8
							3

Whichever method you use, make sure you understand it!

Answer: 223 r3

💡 Tips

- In calculations it is fine to leave a remainder, but in problem solving these need to be presented carefully. You may need to show the remainder, write the remainder as a fraction or a decimal, or round off the answer.

Here's a bit of friendly advice about remainders.

For example: If five pizzas are shared between four people, you wouldn't say each person receives one pizza remainder one. You would say they get $1\frac{1}{4}$ pizzas each.

Or, if a problem asks how many rows of ten can 93 seats be arranged in, the answer is nine. We round the answer and ignore the remainder.

💬 Talk maths

Look at this long division and explain it aloud, saying how each stage was done.

Now try writing down and explaining the steps for this long division:
2878 ÷ 13

> **Remember** that zero divided by anything is ...zero.

				2	2	1	r5
	1	3	2	8	7	8	
(2 × 13 =)		−	2	6			
				2	7		
(2 × 13 =)		−	2	6			
				1	8		
(1 × 13 =)		−	1	3			
					5		

Answer: 221 r5

✔ Check

1. Complete each of these long divisions.

   a.
2	5	5	2	6	4

   b.
1	5	3	8	1	8

2. On squared paper, complete each of these long divisions using a written method.

   a. 338 ÷ 15   b. 4438 ÷ 21   c. 6358 ÷ 18   d. 7318 ÷ 32

⚠ Problems

Brain-teaser A theatre has 2010 seats.
If there are 15 seats per row, how many rows are there? _____

Brain-buster Sixteen people buy a lottery ticket and, altogether, they win £37,468. They agree to share it equally. How much will they each receive, to the nearest 1p? _____

Ordering operations

↺ Recap

Calculations and problems involving more than one operation are called **multi-step**.

You must only do one calculation at a time, and you must do them in the right order!

The right order is division and multiplication first, followed by addition and subtraction, working from left to right.

Look at this calculation:	$25 \div 5 + 3 \times 7 - 6 \times 4$
Division first ($25 \div 5 = 5$)	$5 + 3 \times 7 - 6 \times 4$
Multiplication next ($3 \times 7 = 21$)	$5 + 21 - 6 \times 4$
And another multiplication ($6 \times 4 = 24$)	$5 + 21 - 24$
Then addition ($5 + 21 = 26$)	$26 - 24$
And last subtraction ($26 - 24 = 2$)	Answer $= 2$

📄 Revise

Brackets make a *big* difference.

You can control the order in which calculations are done by using brackets. Calculations inside brackets come first. Look at this example:

$18 - 3 \times 5 = 18 - 15 = 3$ But $(18 - 3) \times 5 = 15 \times 5 = 75$

Or this one:

$21 \div 3 + 4 = 7 + 4 = 11$ But $21 \div (3 + 4) = 21 \div 7 = 3$

💡 Tips

Here's a top tip to keep your maths in order.

- If you understand this you are ready for **BIDMAS**:
  Brackets
  Indices (such as square and cube numbers)
  Division ⎤
  Multiplication ⎦ ← Do multiplication and division together in the order they come, left to right.
  Addition ⎤
  Subtraction ⎦ ← Do addition and subtraction together in the order they come, left to right.
- Indices are a bit tricky. They tell us the power of a number, for example, a square number such as 7^2 is 7 to the power of 2; 7^3 is 7 to the power of 3 and so on.

💬 Talk maths

Look at the calculation below. Try inserting a pair of brackets in different places and discuss, with a partner, what answer it gives you. Remember, do only one calculation at a time, and think BIDMAS.

$$24 + 48 \div 8 - 2 \times 5 - 4 =$$

✔ Check

1. Solve these.

   a. $24 \div 2 - 3 \times 4 = $ _____

   b. $23 - 7 \times 2 - 18 \div 6 = $ _____

   c. $3 \times 45 \div 5 = $ _____

2. Now solve these.

   a. $16 \div (3 + 5) = $ _____

   b. $47 - 7 \times (18 \div 6 + 2) = $ _____

   c. $(7 + 8) \div (12 - 9) = $ _____

3. Mark each of these calculations right (✓) or wrong (✗).

   a. $5 \times 3 - 14 \div 2 = 8$ _____

   b. $(25 - 6) \times 10 \div 5 = 38$ _____

   c. $(8 + 6) - 15 \div 5 \times (4 + 3) = 77$ _____

   d. $(3 \times 7 - 45 \div 5) + 22 - 88 \div (5 + 2 \times 3) = 26$ _____

4. Add the missing brackets to complete calculation correctly.

   a. $8 \times 5 + 2 - 3 = 53$

   b. $14 \div 7 + 2 \times 11 - 6 = 12$

   c. $64 - 12 + 5 \times 3 = 37$

⚠ Problems

Brain-teaser The prize for a charity raffle is £20. Tickets cost £2 each. Charlie sells 34 tickets, Georgina sells 17 tickets and Jayden sells 43 tickets. Georgina says they have made a profit of £168.

Is she right? _____

Write the calculation needed to work out the profit. _____

Brain-buster A car showroom sells new cars for £12,000. It also buys second-hand cars for £2,500 and sells them for £7,000. At the end of a week, the car showroom has received £37,500. Explain how many new cars have been sold, and how many second-hand cars have been bought and sold.

Write the calculation then work out the answer. _____

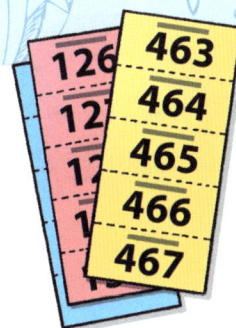

111

Factors, multiples and prime numbers

↺ Recap

A **multiple** is a number that is made by multiplying two numbers.

$$5 \times 7 = 35$$

35 is a multiple of both **5** and **7**.
We can also say that 5 and 7 are factors of 35.

Factors are easy to list in pairs:
The factors of 35 are 1 and 35, 5 and 7.
Factors are the numbers that we multiply together to get multiples.

Prime numbers can only be divided by themselves and one.

DID YOU KNOW?

A *titanic* prime is a prime number that has over 1000 digits!

Remember, 1 is not a prime number, and 2 is the only even prime number.

📝 Revise

A **common factor** is a factor shared by two or more numbers. For example 7 is a common factor of 14 and 77.

A **common multiple** is a multiple shared by two or more numbers. For example 20 is a common multiple of 2 and 5 (and of 1, 4, 10 and 20).

Factors and multiples are easy if you really know your times tables. Try to learn your primes up to 100.

Prime numbers on a 100-square

1	②	③	4	⑤	6	⑦	8	9	10
⑪	12	⑬	14	15	16	⑰	18	⑲	20
21	22	㉓	24	25	26	27	28	㉙	30
㉛	32	33	34	35	36	㊲	38	39	40
㊶	42	㊸	44	45	46	㊼	48	49	50
51	52	㌤	54	55	56	57	58	㊾	60
㊽	62	63	64	65	66	㊿	68	69	70
⑺	72	⑺	74	75	76	77	78	⑺	80
81	82	⑻	84	85	86	87	88	⑻	90
91	92	93	94	95	96	㋆	98	99	100

💡 Tips

- Remember that factors always come in pairs. It can help to list them in pairs too, for example:

  Look at 96 (it has the most factors for any number under 100, 12 altogether):

  96 $= 1 \times 96, \ 2 \times 48, \ 3 \times 32, \ 4 \times 24, \ 6 \times 16, \ 8 \times 12$

💬 Talk maths

Play *True or False* with a partner. Spend ten minutes writing down a collection of facts about factors, multiples and primes, and then take turns challenging your partner to decide if your facts are true or false.

If you give false facts you must know what the true answer should be.

24 has eight factors (True: 1, 2, 3, 4, 6, 8, 12, 24)

100 is a common multiple of 4, 5 and 6 (False: 100 is not a multiple of 6)

38 has two prime factors (True: 2 and 19 are both prime numbers)

✔ Check

1. What are the common factors of 12 and 20? _____

2. What are the common factors of 30 and 50? _____

3. Write three common multiples of 3 and 5. _____

4. What is the lowest common multiple of 2, 5 and 7? _____

5. 30 has three prime factors. What are they? _____

6. What is the largest number between 1 and 100 that has two prime factors? _____

⚠ Problems

Brain-teaser David says, "2 is a prime number and 19 is a prime number. 2 × 19 = 38, so 38 must be a prime number too." Can you explain why David has made a mistake?

_____

_____

_____

Brain-buster Find the highest factor that is shared by 96 and 150.

_____

DID YOU KNOW?

The highest factor that is shared by two numbers is called the highest common factor, or HCF.

Simplifying fractions

One half is one out of *two* equal parts!

↻ Recap

Fractions show proportions of a whole.

They have a **numerator** on the top, and a **denominator** on the bottom.

numerator ⟶ $\dfrac{1}{2}$ ⟵ denominator

📝 Revise

We usually simplify fractions to make them easier to understand.

$\frac{250}{500}$ is the same as $\frac{1}{2}$

It's obvious which one is easier to read and understand.

To simplify fractions you must understand factors.

Look at the dots below.

These statements are true:

Three out of 12 are red.
One in every four is red.

So, $\frac{3}{12}$ is the same as $\frac{1}{4}$.

We say that the fraction has been **simplified**.

We can also simplify fractions using common factors.

To simplify $\frac{24}{30}$ we can separate each number into suitable factor pairs:

Factors of 24 = 1 × 24, 2 × 12, 3 × 8, 4 × 6

Factors of 30 = 1 × 30, 2 × 15, 3 × 10, 5 × 6

Six is the highest common factor of both 24 and 30. Therefore...

$$\frac{24}{30} = \frac{4 \times 6}{5 \times 6}$$

$$\frac{24 \div 6}{30 \div 6} = \frac{4}{5}$$

💡 Tips

Simple tips for simplifying fractions!

- If you can't spot the highest common factor, look for a lower common factor for the numerator and the denominator and divide the numerator and the denominator to simplify and keep going until you get to the smallest number for example:

$$\frac{18 \div 2}{48 \div 2} = \frac{9 \div 3}{24 \div 3} = \frac{3}{8}$$

Or, just divide 18 and 48 by 6!

Talk maths

Play *Big Bang Bong*.

Any number of people can play. You will each need a pencil and paper.

Take turns to call out a fraction (such as, twelve fifteenths). Everyone must write down the fraction in numerator and denominator form, and then it is a race to simplify the fraction as much as possible.

The first person to simplify the fraction must shout bing! Everyone must then agree that they are right.

If they have made a mistake, the first person to spot and correct it shouts bang!

If a fraction has been suggested that cannot be simplified (such as seven sixteenths), the first person to realise this must shout bong!

✔ Check

1. **Write the highest common factor of each pair of numbers.**

   a. 6 and 10 = _____

   b. 15 and 24 = _____

   c. 45 and 17 = _____

   d. 100 and 40 = _____

   e. 30 and 300 = _____

   f. 11 and 88 = _____

2. **Say if these simplifications are true or false.**

   a. $\frac{43}{86} = \frac{1}{2}$ _____

   b. $\frac{12}{60} = \frac{1}{5}$ _____

   c. $\frac{21}{49} = \frac{3}{8}$ _____

   d. $\frac{64}{100} = \frac{16}{25}$ _____

3. **Simplify these fractions.**

   a. $\frac{6}{8} =$ _____

   b. $\frac{15}{20} =$ _____

   c. $\frac{24}{32} =$ _____

   d. $\frac{75}{100} =$ _____

   e. $\frac{36}{80} =$ _____

   f. $\frac{45}{72} =$ _____

   g. $\frac{128}{300} =$ _____

   h. $\frac{64}{200} =$ _____

⚠ Problems

Brain-teaser 128 out of 400 children have school dinners.

Write this as a fraction in its simplest form. _____

Brain-buster What fraction of the children do not have school dinners?

Write the answer in its simplest form. _____

Comparing and ordering fractions

↻ Recap

We can compare and order fractions by giving them the same denominators. To do this we must understand **equivalent fractions**.

This rectangle has been cut into eight equal pieces, or eighths.

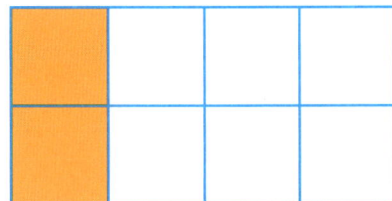

$$\frac{2}{8} = \frac{1}{4}$$ because we have divided the numerator and denominator by 2.

Two eighths is *equivalent* to one quarter because they are the same proportion of the whole.

We can check this by changing either one of them:

$$\frac{2 \div 2 = 1}{8 \div 2 = 4} \qquad \frac{1 \times 2 = 2}{4 \times 2 = 8}$$

> When simplifying a fraction, whatever you do to the numerator, you must do the same to the denominator.

📄 Revise

To compare and order fractions, we must give them the same denominator. Which is bigger, $\frac{2}{5}$ or $\frac{1}{4}$?

We need to find the lowest common multiple which is 20 for 4 and 5, so we must convert each fraction into twentieths.

$$\frac{2 \times 4 = 8}{5 \times 4 = 20} \qquad \frac{1 \times 5 = 5}{4 \times 5 = 20}$$ So, $\frac{2}{5}$ is bigger than $\frac{1}{4}$.

Let's try something harder. Which of these fractions is bigger, $\frac{7}{8}$ or $\frac{17}{20}$ The lowest common multiple for 8 and 20 is 40.

$$\frac{7 \times 5 = 35}{8 \times 5 = 40} \qquad \frac{17 \times 2 = 34}{20 \times 2 = 40}$$ So, $\frac{7}{8}$ is bigger than $\frac{17}{20}$.

💡 Tips

- Remember, when we give each fraction the same denominator, it is called a **common denominator**.
- To compare any number of fractions, you need to give each fraction the same common denominator by finding the **lowest common multiple**. Look at page 112 if you are not certain.
- Remember, > means **is bigger than**, and < means **is smaller than**.

Try this with improper fractions, where the numerator is bigger. The same rules apply!

Talk maths

Write down a selection of fractions, making sure each one has a different numerator and denominator, such as $\frac{3}{7}$ $\frac{2}{3}$ $\frac{5}{8}$ $\frac{1}{9}$ $\frac{4}{6}$.

Next, choose any pair of fractions and change them to give them the same denominator. Then make a statement about them, such as:

$\frac{3}{7}$ and $\frac{2}{3}$ have a common denominator of 21.

$\frac{3}{7} = \frac{9}{21}$ and $\frac{2}{3} = \frac{14}{21}$ so $\frac{2}{3} > \frac{3}{7}$.

✔ Check

1. Change each fraction to give it a denominator of 30.

   a. $\frac{1}{2} =$ _____
   b. $\frac{2}{3} =$ _____
   c. $\frac{3}{5} =$ _____
   d. $\frac{5}{6} =$ _____

2. Insert the correct sign, =, < or >.

   a. $1\frac{1}{2}$ _____ $1\frac{3}{6}$
   b. $3\frac{3}{4}$ _____ $3\frac{2}{3}$
   c. $\frac{20}{6}$ _____ $\frac{13}{4}$
   d. $\frac{12}{5}$ _____ $\frac{15}{6}$

3. True or false?

   a. $\frac{3}{7} > \frac{1}{3}$ _____
   b. $\frac{15}{9} > \frac{7}{5}$ _____
   c. $\frac{7}{11} > \frac{13}{20}$ _____

4. Arrange these fractions in order, smallest to largest. Place a less than sign (<) between each one.

   a. $\frac{3}{4}, \frac{5}{8}, \frac{2}{3}$: _____
   b. $\frac{4}{9}, \frac{3}{7}, \frac{1}{3}$: _____
   c. $\frac{13}{24}, \frac{5}{9}, \frac{7}{12}$: _____

⚠ Problems

Brain-teaser Eva's mum has some money in her purse. She says that Eva can have a fraction of it. She offers Eva $\frac{3}{8}$ or $\frac{7}{20}$ of the money.

Which fraction will give Eva more money? _____

Brain-buster In a survey, some children were asked which pets they owned. $\frac{2}{7}$ of the children owned dogs and $\frac{3}{12}$ owned cats. The others owned no pets. Arrange the three sets of children in order, showing the fraction of each.

_____ < _____ < _____

Adding and subtracting fractions

↺ Recap

To add and subtract fractions, they must have the same denominator.

To add $\frac{1}{2}$ and $\frac{1}{3}$, first find the lowest common denominator ($2 \times 3 = 6$).

Next, convert each fraction to give it a denominator of 6.

$$\frac{1 \times 3 = 3}{2 \times 3 = 6} \qquad \frac{1 \times 2 = 2}{3 \times 2 = 6}$$

Then, add the new fractions:

$$\frac{3}{6} + \frac{2}{6} = \frac{5}{6}$$

> **And you must only add the numerators!**
> $\frac{3}{12} + \frac{4}{12} = \frac{7}{12}$

Taking away is exactly the same – you only subtract the numerators.

$$\frac{7}{10} - \frac{3}{10} = \frac{4}{10}$$

📝 Revise

The common denominator will usually be the lowest common multiple of all the fractions involved.

If one denominator is a multiple of the other, you only need to change one. For example:

$$\frac{3}{5} + \frac{1}{10} = \frac{6}{10} + \frac{1}{10} = \frac{7}{10}$$

Sometimes you will need to think more, for example:

$\frac{3}{5} + \frac{1}{8}$ 40 is the lowest common multiple of 5 and 8.

$$\frac{3}{5} = \frac{24}{40} \text{ and } \frac{1}{8} = \frac{5}{40}$$

💡 Tips

> **Here's how to add mixed numbers and improper fractions.**

- There are two ways to deal with improper fractions and mixed numbers.

   1. Add the whole numbers and the fractions separately.

   $1\frac{1}{3} + 3\frac{5}{6}$

   $= 1 + 3 + \frac{1}{3} + \frac{5}{6}$

   $= 4 + \frac{2}{6} + \frac{5}{6}$

   $= 4\frac{7}{6} = 5\frac{1}{6}$

   2. Use improper fractions.

   $1\frac{1}{3} + 3\frac{5}{6}$

   $= \frac{4}{3} + \frac{23}{6}$

   $= \frac{8}{6} + \frac{23}{6}$

   $= \frac{31}{6} = 5\frac{1}{6}$

   It works for subtraction too!

Talk maths

$$\frac{1}{2} \quad \frac{3}{4} \quad \frac{1}{3} \quad \frac{7}{12} \quad \frac{1}{6} \quad \frac{3}{10}$$

$$\frac{2}{3} \quad \frac{5}{9} \quad \frac{4}{5} \quad \frac{5}{8} \quad \frac{5}{6} \quad \frac{3}{4}$$

Work with a partner and challenge them to add and subtract fractions. You can *only* say fractions that have one denominator that is a multiple of the other. Use the fractions in the box, or make up some of your own, for example:

Add one third and five sixths

$(\frac{1}{3} + \frac{5}{6} = \frac{2}{6} + \frac{5}{6} = \frac{7}{6} = 1\frac{1}{6})$

Challenge your partner to work it out then read their answer to you.

One third plus five sixths equals seven sixths, or one and one sixth.

✔ Check

1. Add these fractions.

   a. $\frac{1}{6} + \frac{2}{3} =$ _____

   b. $\frac{2}{5} + \frac{3}{10} =$ _____

   c. $\frac{1}{4} + \frac{1}{8} + \frac{1}{2} =$ _____

2. Subtract these fractions.

   a. $\frac{5}{8} - \frac{1}{2} =$ _____

   b. $\frac{7}{9} - \frac{1}{3} =$ _____

   c. $\frac{7}{12} - \frac{2}{5} =$ _____

3. Insert the missing sign (+ or –).

   a. $\frac{1}{2} \rule{1cm}{0.4pt} \frac{1}{4} = \frac{3}{4}$

   b. $\frac{1}{2} \rule{1cm}{0.4pt} \frac{1}{3} = \frac{1}{6}$

   c. $\frac{1}{2} \rule{1cm}{0.4pt} \frac{2}{5} = \frac{9}{10}$

   d. $\frac{2}{7} \rule{1cm}{0.4pt} \frac{1}{6} = \frac{5}{42}$

   e. $\frac{7}{10} \rule{1cm}{0.4pt} \frac{1}{4} = \frac{9}{20}$

   f. $\frac{3}{8} \rule{1cm}{0.4pt} \frac{1}{12} = \frac{11}{24}$

4. Complete these calculations.

   a. $\frac{5}{2} + \frac{7}{4} =$ _____

   b. $2\frac{1}{2} - 1\frac{1}{4} =$ _____

   c. $\frac{10}{3} - \frac{11}{5} =$ _____

   d. $2\frac{2}{3} + 1\frac{4}{5} =$ _____

⚠ Problems

Brain-teaser Emma and Tom buy a pizza. If Emma eats $\frac{1}{2}$ of it and Tom eats $\frac{1}{3}$,

how much pizza is left over? _____

Brain-buster Richard and Amy have some popcorn. Richard eats three sevenths of it and Amy eats four elevenths of it.

How much popcorn is left? _____

Multiplying fractions

Remember, multiplication works in any order: $\frac{1}{2} \times 24$ is the same as $24 \times \frac{1}{2}$.

↻ Recap

We can multiply whole numbers by fractions.

When multiplying by a fraction we use the word **of**.

- $\frac{1}{2}$ of 10 = 5.
- One quarter of 12 is 3.
- $\frac{1}{3}$ of 9 is 3.

Revise

We can also multiply fractions by other fractions.

Watch carefully: when we multiply fractions together we multiply the numerators with each other *and* we multiply the denominators with each other.

$$\frac{1}{2} \times \frac{3}{4} = \frac{1 \times 3}{2 \times 4} = \frac{3}{8}$$

Look at the circle opposite. Can you see how half of three quarters equals three-eighths?

Let's try something harder:

$$\frac{5}{6} \times \frac{2}{3} = \frac{5 \times 2}{6 \times 3} = \frac{10}{18}$$ (we can simplify this to $\frac{5}{9}$)

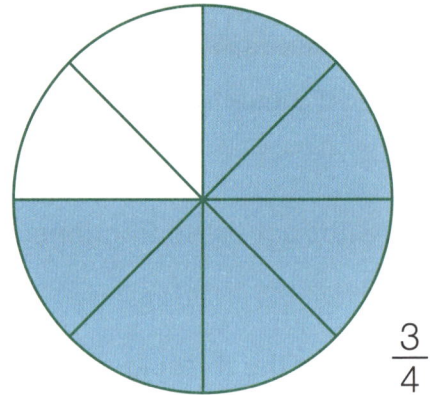

$\frac{3}{4}$

All whole numbers can be written as fractions with a denominator of 1.

So, $5 \times \frac{3}{8}$ is the same as saying $\frac{5}{1} \times \frac{3}{8} = \frac{15}{8}$ (or $1\frac{7}{8}$).

💡 Tips

This trick might save you time, but only use it if you understand it!

- Look at this calculation:

$$\frac{2}{3} \times \frac{3}{5} = \frac{2 \times 3}{3 \times 5} = \frac{6}{15} = \frac{2}{5}$$

We didn't really need to do a calculation because the three on the top cancels out with the three on the bottom (3 ÷ 3 = 1).

- Can you see the quick way to solve this calculation?

$$\frac{3}{7} \times \frac{7}{9} = \frac{3 \times 7}{7 \times 9} = \frac{3}{9} = \frac{1}{3}$$

- Remember to simplify fractions as much as possible.

Talk maths

Choose any two fractions from the examples in the box. Read them aloud as a multiplication. Try solving the problem mentally, explaining your answer.

$\frac{7}{10}$	$\frac{5}{8}$	$\frac{5}{6}$	$\frac{2}{7}$	$\frac{4}{5}$	$\frac{1}{4}$	$\frac{2}{3}$	$\frac{1}{2}$

$\frac{7}{10} \times \frac{4}{5} = \frac{28}{50}$ because **7 × 4 = 28**, and **10 × 5 = 50**.

Also, 2 is a common factor of 28 and 50, so we can simplify to $\frac{14}{25}$.

✔ Check

1. Complete these multiplications.

   a. $\frac{1}{2}$ of 20 = _____

   b. $\frac{1}{4}$ of 24 = _____

   c. $\frac{3}{4}$ of 24 = _____

   d. $\frac{2}{5} \times 25$ = _____

   e. $\frac{5}{6}$ of 30 = _____

   f. $\frac{2}{3} \times 39$ = _____

2. Write these answers as mixed numbers.

   a. $14 \times \frac{1}{4}$ = _____

   b. $25 \times \frac{1}{2}$ = _____

   c. $40 \times \frac{1}{3}$ = _____

   d. $14 \times \frac{3}{7}$ = _____

   e. $12 \times \frac{3}{5}$ = _____

   f. $100 \times \frac{1}{6}$ = _____

3. Multiply these fractions.

   a. $\frac{1}{2} \times \frac{1}{3}$ = _____

   b. $\frac{2}{5} \times \frac{3}{4}$ = _____

   c. $\frac{3}{8} \times \frac{8}{9}$ = _____

   d. $\frac{5}{6} \times \frac{4}{5}$ = _____

   e. $\frac{2}{3} \times \frac{5}{8}$ = _____

   f. $\frac{10}{7} \times \frac{4}{5}$ = _____

⚠ Problems

Brain-teaser Tinashe usually takes ten and a half minutes to run one lap of the park. In her roller skates she can do the same lap in half this time. How long will it take her in roller skates?

_____

Brain-buster A second is $\frac{1}{60}$ of a minute, and a minute is $\frac{1}{60}$ of an hour.

What is a second as a fraction of an hour? _____

Dividing fractions

↻ Recap

When multiplying by a fraction we multiply the numerators together, and we multiply the denominators together.

$$\frac{1}{5} \times \frac{3}{4} = \frac{1 \times 3}{5 \times 4} = \frac{3}{20}$$

Revise

Just as we can multiply fractions, we can also divide fractions. Look at the circle opposite. Half has been shaded.

If we divide the shaded half in two we get quarters.

So: $\frac{1}{2} \div 2 = \frac{1}{4}$.

Remember $\frac{1}{2} \div \frac{1}{2} = \frac{1}{4}$. So, dividing by 2 is the same as multiplying by $\frac{1}{2}$.

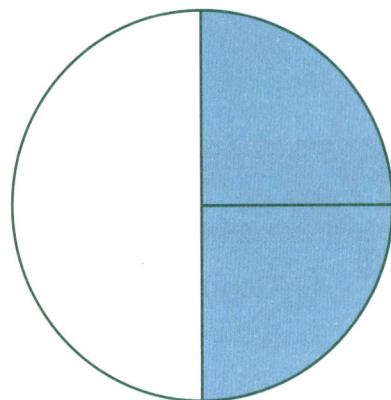

Now try this one: $\frac{1}{4} \div 3$

This is the same as saying $\frac{1}{4} \div \frac{1}{3} = \frac{1}{12}$.

Try drawing a circle and dividing it into fractions to prove this.

♀ Tips

- Dividing fractions is tricky.
  But remember that dividing by a whole number is the same as multiplying by one over that number, such as:

  $\frac{2}{3} \div 5$ is the same as $\frac{2}{3} \div \frac{5}{1}$ which is the same as $\frac{2}{3} \times \frac{1}{5} = \frac{2}{15}$.

  So, $\frac{2}{3} \div 5 = \frac{2}{15}$.

🗨 Talk maths

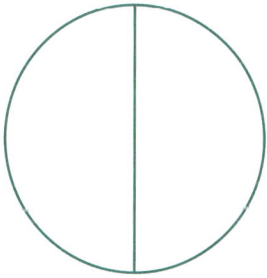

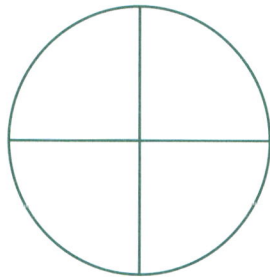

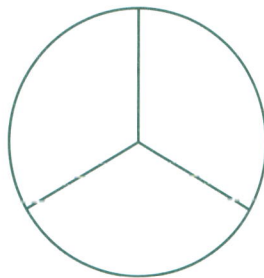

These circles have been divided into halves, quarters and thirds. Use them to help you discuss dividing simple fractions by whole numbers.

A half divided by three equals one sixth.

A quarter divided by two equals one eighth.

A third divided by five equals one fifteenth.

✔ Check

1. Mark each calculation right (✓) or wrong (✗).

a. $\frac{1}{2} \div 3 = \frac{1}{6}$ _____

b. $\frac{1}{4} \div 2 = \frac{1}{8}$ _____

c. $\frac{1}{3} \div 3 = \frac{1}{6}$ _____

d. $\frac{2}{5} \div 4 = \frac{1}{10}$ _____

e. $\frac{3}{4} \div 2 = \frac{1}{2}$ _____

f. $\frac{6}{3} \div 4 = \frac{1}{2}$ _____

2. Complete these divisions.

a. $\frac{1}{2} \div 2 =$ _____

b. $\frac{1}{4} \div 3 =$ _____

c. $\frac{1}{3} \div 5 =$ _____

d. $\frac{2}{3} \div 4 =$ _____

e. $\frac{3}{4} \div 4 =$ _____

f. $\frac{2}{3} \div 20 =$ _____

⚠ Problems

Brain-teaser Jem shares half a cake between seven people.

What fraction of the whole cake will they each receive? _____

Brain-buster A teacher has three tenths of a sheet of stickers left and wants to share them equally among her class of 24 children.

What fraction of the sheet of stickers will each child receive? _____

If a full sheet holds 240 stickers, how many will each child receive? _____

Decimal equivalents

↻ Recap

A proper fraction is a proportion of one whole.

$\frac{1}{4}, \frac{1}{3}, \frac{1}{2}, \frac{2}{3}, \frac{3}{4}$ are all proper fractions.

A fraction is a numerator divided by a denominator, such as:

$\frac{1}{2}$ **is 1 divided by 2, so $\frac{1}{2}$ = 0.5**

You need to learn these common fractions and their decimal equivalents:

Fraction	$\frac{1}{2}$	$\frac{1}{4}$	$\frac{3}{4}$	$\frac{1}{5}$	$\frac{1}{10}$
Decimal	0.5	0.25	0.75	0.2	0.1

📋 Revise

Any fraction can be written as a decimal.
If you need to calculate the decimal equivalent of a fraction, just do a short division.

$$\frac{1}{4} = 4\overline{)1.{}^1 0\,{}^2 0}\quad 0.25$$

$$\frac{3}{8} = 8\overline{)3.{}^3 0\,{}^6 0\,{}^4 0}\quad 0.375$$

> **Notice that a whole number can be written with zeros in the decimal places.**

Remember to keep the decimal point in the right place!

💡 Tips

- Remember that, after a decimal point, the first column is tenths, the second column is hundredths, and the third column is thousandths.
- We read decimals aloud using the numbers zero to nine.
  We say 0.5 is **zero point five**.
  We say 0.75 is **zero point seven five**.
  We say 0.375 is **zero point three seven five**.
  We say 0.666 is **zero point six six six**.

> **Time for some decimal tips!**

💬 Talk maths

You know about fraction equivalents, such as $\frac{2}{4} = \frac{1}{2}$.

Now look at what happens when they are changed to decimals.

$$\frac{2}{4} = 0.5 \qquad \frac{1}{2} = 0.5$$

Because the fractions are equivalent, they both equal 0.5.

Discuss this with an adult, completing this chart as you go.

Fraction	$\frac{2}{8} = \frac{1}{4}$	$\frac{4}{10} = \frac{2}{5}$	$\frac{2}{6} = \frac{1}{3}$	$\frac{6}{8} = \frac{3}{4}$	$\frac{10}{12} = \frac{5}{6}$
Decimal	0.25	0.4			

✔ Check

1. Convert these fractions to decimals.

   a. $\frac{2}{5} =$ _____
   b. $\frac{6}{10} =$ _____
   c. $\frac{3}{8} =$ _____

2. Complete this chart.

Fraction	$\frac{1}{8}$	$\frac{2}{8}$	$\frac{3}{8}$	$\frac{4}{8}$	$\frac{5}{8}$	$\frac{6}{8}$	$\frac{7}{8}$	$\frac{8}{8}$
Decimal	0.125	0.25						

3. Match each fraction to its decimal equivalent.

   $\frac{3}{4}$ $\frac{5}{8}$ $\frac{4}{5}$ $\frac{1}{3}$

   0.625   0.8   0.333   0.75

4. Match each decimal to its fraction equivalent.

   0.166   0.4   0.7   0.125

   $\frac{1}{8}$ $\frac{1}{6}$ $\frac{7}{10}$ $\frac{2}{5}$

⚠ Problems

Brain-teaser Which is more, $\frac{5}{6}$ or 0.8? _____

Brain-buster A bag of popcorn is shared equally between 12 people. Tim says that each person will receive 0.1 of the popcorn. Is he right? Explain your answer.

_____

_____

Decimal places

↻ Recap

A decimal fraction has 10, 100 or 1000 as its denominator, such as $\frac{4}{10}$.

We can say $\frac{4}{10}$ as 4 divided by 10.

When we divide a number by 10, 100 or 1000, we move the numbers to the right.

Fraction name	Fraction	Decimal	Decimal name
seven tenths	$\frac{7}{10}$	0.7	Zero point seven
twenty three hundredths	$\frac{23}{100}$	0.23	Zero point two three
four hundred and thirty five thousandths	$\frac{435}{1000}$	0.435	Zero point four three five

> The place value of each digit changes.

📋 Revise

> Money usually needs to be rounded to two decimal places.

Decimals can have more than three decimal places, but usually we round decimals, just like we round other numbers.

A basic rule for rounding is if the next number is five or more, round up, if not, round down.
0.87 = 0.9 to one decimal place
0.435 = 0.44 to two decimal places
0.2574 = 0.257 to three decimal places

Look at these examples.

Fraction	$\frac{1}{7}$	$\frac{7}{17}$
Decimal	0.142857	0.411764
Rounded to three decimal places	0.143	0.412
Rounded to two decimal places	0.14	0.41
Rounded to one decimal place	0.1	0.4

💡 Tips

- Some decimals have the same number that goes on forever, such as
  $\frac{1}{6} = 0.16666666666666666666666666666666666666$

  We call this a recurring decimal. We usually round these decimals to three decimal places.

  So $\frac{1}{6} = 0.167$ to three decimal places.

- $\frac{1}{3}$ and $\frac{2}{3}$ also make recurring decimals.

- $\frac{1}{3} = 0.333$ to three decimal places. $\frac{2}{3} = 0.667$ to three decimal places.

💬 Talk maths

You will need two or more people.

Think of a fraction with demoninators of 2, 4, 5 or 8 and then use division to calculate the decimal equivalent.

$\frac{2}{5} = 0.4$

Take turns to challenge each other to say the decimal to one, two or three decimal places, checking each other's answers.

✔ Check

1. **Look at these decimals and say how many thousandths, hundredths and tenths each one has.**

   a. 0.375 _____ thousandths _____ hundredths _____ tenths

   b. 0.903 _____ thousandths _____ hundredths _____ tenths

2. **Complete this chart.**

Fraction	Decimal	Rounded to three decimal places	Rounded to two decimal places	Rounded to one decimal place
$\frac{2}{7}$	0.285714			
$\frac{3}{13}$	0.230769			
$\frac{4}{11}$	0.363636			
$\frac{2}{3}$	0.666667			
$\frac{8}{9}$	0.888889			

⚠ Problems

Brain-teaser Jared says that 0.001 rounded to the nearest tenth is 0.1. Is he right? _____

Explain your answer. _____

Brain-buster Explain why $\frac{3}{11}$ is a recurring number, and round it to three decimal places.

_____

Multiplying decimals

↺ Recap

Do you remember what tenths, hundredths and thousands are?
Tenths are bigger than hundredths, and hundredths are bigger than thousandths.

| $0.6 > 0.5$ | $0.431 > 0.429$ | $0.1 > 0.099$ | $0.3 > 0.28$ | $0.515 > 0.4$ |

- There are ten tenths in a whole.
- There are one hundred hundredths in a whole, but ten hundredths in one tenth.
- There are one thousand thousandths in a whole, but ten thousandths in one hundredth.

Ones	Tenths	Hundredths	Thousandths
0 .	1	2	3

one hundred and twenty-three thousandths
= zero point one two three
$\frac{123}{1000} = 0.123$

🗒 Revise

We can multiply any two numbers together, including numbers that are decimals.
For the moment, we will learn how to multiply a decimal by a whole number.
This will come in very handy for solving money problems!

Do you remember how to use formal written methods for multiplication?

```
      3   2   4
  ×       1   3
  ─────────────
      9   7¹  2
  3   2   4   0   +
  ─────────────
  4   2   1   2
      ╷   ╷
```
Answer: 4212

Well, the same method works for decimals.

```
      4 · 1   3
  ×       2   3
  ─────────────
  1   2 . 3   9
  8   2 . 6   0   +
  ─────────────
  9   4 · 9   9
```
Answer: 94.99

> It's all about place value. Just remember to keep the decimal point in the right place.

💡 Tips

- When you multiply a decimal by a whole number, make sure you give your answer as a decimal too, with the same number of decimal places, for example:
  6.35 × 2 = 12.70 or **£1.25 × 4 = £5.00**
  Keeping the zeros helps with checking work later on.

Talk maths

With a friend, investigate multiplying decimals by whole numbers. Use small numbers to see if you can spot any handy patterns, such as

3 × 4 = 12, 0.3 × 4 = 1.2   or   6 × 8 = 48, 6 × 0.8 = 4.8

✔ Check

1. Complete each of these decimal multiplications using a written method.

a.

	0 .	2
×		3

b.

	3 .	3
×		2

c.

	0 .	2	3
×			4

d.

	0 .	3	4
×			6

2. Complete each of these long divisions using a written method.

a. 0.23 × 21

b. 0.45 × 15

c. 0.25 × 25

d. 3.33 × 33

⚠ Problems

Brain-teaser A group of eight friends decide to buy their teacher some flowers.

If they each contribute £1.15, how much will they have? _____

Brain-buster A school trip is going to cost exactly £100. A letter is sent home asking for a donation of £2.65 per child towards the trip. If there are 32 children in the class and they all make the contribution, how much more will the school have to contribute?

_____

Dividing decimals

↺ Recap

Short division

	0	4	2	6	r2
8	3	³4	²1	⁵0	

In short division we carry on the remainder at each stage, but with long division we calculate the remainder at each stage, so that there is less chance of errors.

Long division

				2	2	3	r3
	1	6	3	5	7	1	
		−	3	2	↓		
				3	7		
		−	3	2	↓		
				5	1		
		−	4	8			
					3		

📄 Revise

We can use short and long division for dividing decimals.
Just remember to keep the decimal point in the same place.

	0	.	8	7	
4	3	.	³4	²8	

				1	.	6	4
	1	3	2	1	.	3	2
(13 × 1 →)	−		1	3	↓		↓
				8	.	3	
(13 × 6 →)	−		7	8			↓
				5		2	
(13 × 4 →)	−		5	2			
				0		0	

✔ Check

1. Complete each of these short divisions of decimals.

a.

3	0	.	3	9

b.

2	0	.	5	4

c.

4	0	.	9	6

2. On paper, complete these long divisions of decimals using a written method.

   a. $0.6 \div 15$ b. $7.04 \div 32$

   c. $3.30 \div 22$ d. $77.4 \div 15$

⚠ Problems

Brain-teaser Aysha has a brother and a sister. Their mum gives them £10.44 pocket money, to share equally between the three of them. How much will they each get?

_____

Percentage equivalents

Decimal fractions can be called percentages.

↺ Recap

$\frac{65}{100}$ is a decimal fraction.
We can say **65 over 100** or **65 out of 100**.

Per cent means **parts of a hundred** or **out of 100**.
Look at the 100 grid. 65 out of the 100
squares are shaded, this is 65%.

$$0.65 = \frac{65}{100} = 65\%$$

📄 Revise

It is easy to find the equivalents of simple fractions and decimals.

We can use our knowledge of decimal places and rounding to help us find trickier equivalents.

Fraction	$\frac{1}{2}$	$\frac{1}{4}$	$\frac{1}{10}$	$\frac{1}{5}$	$\frac{3}{4}$	$\frac{1}{1}$
Decimal	0.5	0.25	0.1	0.2	0.75	1.0
Per cent	50%	25%	10%	20%	75%	100%

$$\frac{3}{8} = 0.375 = 37.5\% \qquad \frac{5}{6} = 0.833 = 83.3\%$$

✔ Check

1. Complete the chart.

Percentage	Decimal	Fraction
33.3%		
	0.125	
		$\frac{2}{5}$
	0.85	
		$\frac{7}{8}$

⚠ Problems

Brain-teaser 12 out of 30 children have blond hair. What is that as a percentage? _____

Ratio and proportion: numbers

And, of course, three out of four of the squares are red.

↻ Recap

A fraction shows us one number compared to a whole.
In the shape opposite, one out of four of the squares is blue.

Proportion is the fraction of a whole.
For this shape, the proportion of blue squares is one in four, or one out of four. And the proportion of red squares is three in four, or three out of four.

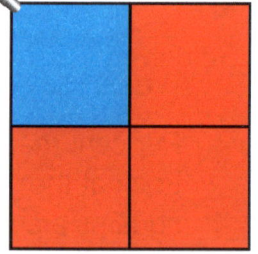

Ratio is different, because it compares amounts.
For the shape above, the ratio of blue squares to red squares is 1 to 3, or 1:3.

📄 Revise

Look at these examples.

In total there are 100 animals on a farm.
There are two dogs, three cats, five rabbits, 20 cows, 30 sheep and 40 chickens.

Proportion

The proportion of dogs is two out of 100 animals. As a fraction this is $\frac{2}{100}$ or $\frac{1}{50}$.

The proportion of rabbits is $\frac{5}{100}$ or $\frac{1}{20}$.
One in every 20 animals is a rabbit.

Ratio

The ratio of dogs to cows is 2:20.
This can be simplified to 1:10.
There are ten cows for every dog.

The ratio of cows to chickens is 20:40.
This can be simplified to 1:2.
For every cow there are two chickens.

💡 Tips

- Proportion is a fraction of the whole; ratio compares different amounts.
- One in every five adults play computer games (so four out of five do not play).
  As a *proportion* this is one out of five, or $\frac{1}{5}$.
  But the *ratio* of adults who do play to adults who don't play computer games is 1:4.

Talk maths

A wall is covered with 100 tiles.

Ten are black, 20 are white, 15 are red, 15 are yellow and 40 are blue.

Work with a friend to agree on some proportion and ratio statements about the tiles.

Remember to write the ratio in the simplest form.

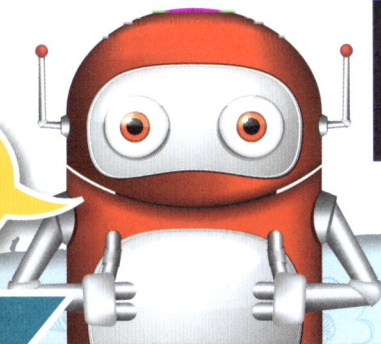

✔ Check

1. **Write the proportion of black squares in each pattern.**

   a. _____   b. _____   c. _____

2. **Look at this pattern and write the ratios.**

   a. Blue to red ____:____   b. Red to green ____:____   c. Yellow to green ____:____

⚠ Problems

Brain-teaser In a class of 30 pupils, six of the class can speak two languages.

a. What proportion of the class can speak two languages? _____

b. What is the ratio of dual-language to single-language speakers? _____

Brain-buster A recipe for a fruit pie says to add blackberries and blueberries in the ratio 3:4.

a. If Hana has 15 blackberries, how many blueberries will she need? _____

b. What proportion of the berries will be blueberries? _____

Ratio and proportion: percentages

↻ Recap

> And the proportion of green triangles is 2 in 3, or 2 out of 3.

Proportion is the fraction of a whole. For this shape, the proportion of yellow triangles is one in three, or one out of three.

Ratio compares amounts. For this shape, the ratio of yellow to green triangles is one to two, or 1:2.

> And the ratio of *green to yellow* triangles is two to one or 2:1.

📄 Revise

> That's easy. What if there were only 50 children?

Percentages are a type of proportion. They represent an amount out of 100.

35 children **out of 100** have packed lunches, which is $\frac{35}{100}$ or 35%.

If 17 children out of 50 have brown eyes, as a proportion it is $\frac{17}{50}$.

Percentages must be out of 100, so we must adjust the fraction.

$$\frac{17}{50} = \frac{34}{100} \text{ so 34\% have brown eyes.}$$

Remember that 100% is everything, so, if 34% of the children have brown eyes, 66% do not, because 34% + 66% = 100%.

💡 Tips

- When calculating percentages, choose the order of calculations you find easier, for example, to find 26% of 360:

  - You can either find 25% ($\frac{1}{4}$) of 360 = 90, plus 1% of 360 = 3.6.

    360 × 25 = 90 **plus** 360 ÷ 1 = 3.6
    or 90 + 3.6 = 93.6

  - Or you can do 26 × 360, then divide by 100.

    26 × 360 = 9360
    9360 ÷ 100 = 93.6

Talk maths

You will need a pack of playing cards with the picture cards removed.
This will leave 40 cards, 1–10 in each suit of clubs, diamonds, spades and hearts.

Sort the pack in different ways and then make statements of proportion, ratio and percentage, such as:
One in four cards is a diamond.
The ratio of diamonds to other cards is 1:3.
25% of the pack is diamonds.

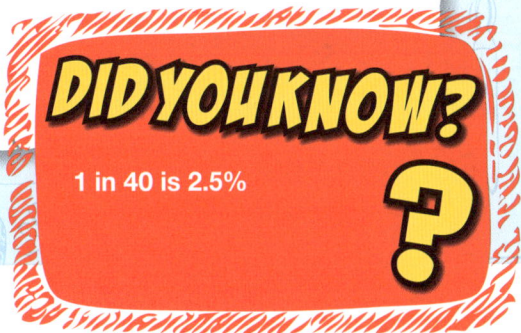

DID YOU KNOW?

1 in 40 is 2.5%

✔ Check

1. Write these proportions as a percentage.

   a. 1 in 4 = _____   **b.** 7 in 10 = _____   **c.** 2 in 5 = _____   **d.** 3 in 8 = _____

2. Write these percentages as a proportion in their simplest form.

   a. 25% = _____   **b.** 40% = _____   **c.** 26% = _____   **d.** 87.5% = _____

3. Calculate these percentages.

   a. 25% of 200 = _____   **b.** 50% of 1 = _____   **c.** 10% of 624 = _____

   d. 95% of 300 = _____   **e.** 60% of 24 = _____   **f.** 15% of 360 = _____

4. Explain what each of these mathematical terms mean.

   a. Percentage: _____

   b. Proportion: _____

   c. Ratio: _____

⚠ Problems

Brain-teaser In a traffic survey, children counted 220 cars. 25% were driving over the speed limit.

How many cars were driving too fast? _____

Brain-buster The percentage of homes in the UK where a dog is kept as a pet is 18%.

If there are 42 million homes in total, how many of these will keep a dog? _____

Scale factors

And the ratio of yellow to red beads is five to one, or 5:1.

↻ Recap

Ratio compares amounts.
For every red bead there are five yellow beads.
The ratio of red to yellow beads is 1:5.

📋 Revise

People often draw to scale. This means changing the proportion of what is drawn.

We have to draw things to scale to fit our drawings on the paper!

Scale is usually shown as a ratio.
The brown line is four times longer than the green line.
The scale of **green**:**brown** is **1:4**.

Jez is 125cm tall.
His brother draws a picture of her using a scale of 1:10.
His drawing will be 12.5cm tall.

The scale of this map is 1:100,000.
Every 1cm on the map represents
100,000cm (or 1km) in real life.

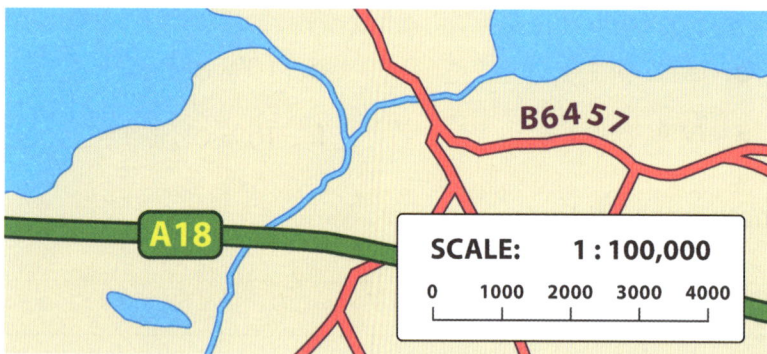

B6457

A18

SCALE: 1 : 100,000

0   1000   2000   3000   4000

💡 Tips

Scaling up and down is easy when you follow my tips!

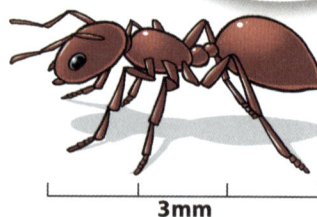

- Remember that scale can work the other way round too. If you want to draw an insect, it's easier to enlarge it. So, if an ant is 3mm long an enlargement of 50:1 would give a drawing of 50 × 3 = 150mm, or 15cm.

3mm

Talk maths

Or just draw their hand – remember to measure it before you start.

You will need a sheet of paper, a pencil and a ruler. Measure these objects, and then try to draw an enlargement of each object, using a scale of 5:1. Take your drawings and explain them to an adult. To finish, try a friend or an adult at a scale of 1:10.

✔ Check

1. **This line is 4cm long.**

    How long would these enlargements be?

    a. 2:1 _____   **b.** 5:1 _____   **c.** 10:1 _____

2. **This square has a side of 1cm.**

    Complete this chart for different scale enlargements.

Scale of enlargement	Side length	Area
5:1		
10:1		
25:1		

3. **A table is 1m high.**

    What height would models be if they were made to these scales?

    a. 1:2 _____   **b.** 1:5 _____   **c.** 1:20 _____

⚠ Problems

Brain-teaser A model of a house is made to a scale of 1:25.

If the model is 22cm high, what height is the actual house? _____

Brain-buster Anita makes a sculpture of a mouse.
The actual mouse is 8cm high. The sculpture is 60cm high.

What is the scale of the enlargement? _____

Using simple formulae

↻ Recap

If we need to calculate the perimeter or area of a regular shape, we can use a formula.

For the rectangle, we can say,
Area equals length multiplied by width.
In a formula, we can use a letter for each part.
So, **area equals length multiplied by width**
becomes **$A = l \times w$**.

A = area l = length w = width

←——— Length (l) ———→

↑ Width (w) ↓

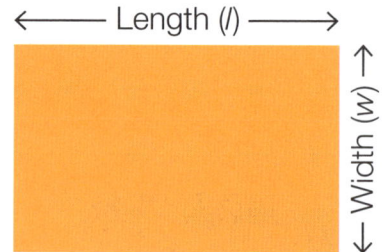

🗒 Revise

←——— Length 4cm ———→

↑ Width 3cm ↓

In formulae, we can drop the multiplication sign. If a letter and a number, or two letters, are together, it means that they are being multiplied.

The area of a rectangle is **$A = lw$**.
For the red rectangle, **$A = 4 \times 3 = 12cm^2$**

For a rectangle that is 7m long and 2m wide: $A = 7 \times 2 = 14m^2$.
For a rectangular field that is 90m long and 30m wide:
$A = 90 \times 30 = 2700m^2$.

Notice that area has square units. It is shown with this symbol ^{2}.

Perimeter is the distance around a shape.
For a rectangle $P = l + w + l + w$ or, $P = 2l + 2w$.
Remember, multiplication before addition.
For the red rectangle, $P = 2 \times 4 + 2 \times 3 = 14cm$.

You can use other formulas in the same way. Just replace the letters with the numbers.

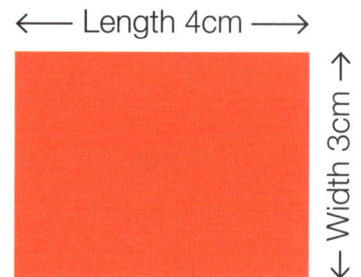

The great thing about a formula is that you can use it again and again. The letters always stay the same but they represent different numbers.

💡 Tips

● Be sure to get your units right. Formulae are used to calculate all sorts of things: distance, area, temperature, weight, volume, and so on. You must be sure to keep everything in the same units.

● If you are calculating with different units, you must convert one unit to the other first: you must multiply centimetre by centimetre, add grams to grams, and so on.

🗨 Talk maths

Try inventing your own simple formulae, and then test them on an adult, for example:

- Some new houses are being built.
  If every house has seven windows, a formula for windows is:
  $w = 7h$, where h = the number of houses, and w = the number of windows.
- How about cars? You need five tyres per car.
- Or currant buns? There are 24 currants per bun!

> If there are six houses there must be 42 windows!

> If there are 100 houses, there will be 700 windows!

✔ Check

1. Complete the chart for perimeters and areas of rectangles.

Length	Width	Perimeter	Area
5cm	2cm		
5m	4m		
7km	1.5km		
3.2m	2.3m		

2. Use this formula to complete the chart:
$h = 3f + 8$

h					
f	1	2	4	9	100

⚠ Problems

Brain-teaser Beth wants to change some pounds to dollars. The formula for calculating the amount of pounds she receives is $ = 1.67 × £. £ is the amount of pounds Beth has and $ is the dollars she will receive. (1.67 is called the exchange rate.)

If Beth has £200 to change, how many dollars will she receive? _____

Brain-buster Here is the formula for changing degrees Fahrenheit to degrees Celsius:

$C = \frac{5}{9} × (F - 32)$

Use the formula to complete this chart.

Fahrenheit	32°	104°	212°
Celsius			

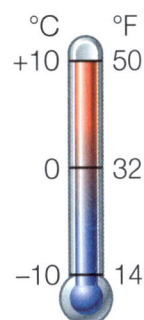

°C °F
+10 50
0 32
−10 14

Missing numbers

↻ Recap

Sometimes equations have missing numbers.

4 + ⬤ = 12

Easy! The missing number is 8.

4.3 − ⬤ = 3.1

Not so easy! The missing number is...1.2.

🗐 Revise

For harder problems, it can help to put a letter in the place of the missing number.

⬤ − 9 = 23

$h - 9 = 23$

$h - 9 + 9 = 23 + 9$

$h = 32$

Now try this one: 3 × ⬤ = 30

And this one: 4 + 2 × ⬤ = 30

Because the equation must balance, you must add the same amount to each side of the equals sign. Look...

💡 Tips

The numbers might be missing, but the tip isn't...

- Don't forget that missing numbers could be negative numbers or decimals. Can you see the answers for these two?

  ⬤ + 3 = 2   3.1 + ⬤ = 7.5

  (missing number = −1)   (missing number = 4.4)

DID YOU KNOW?

An equation must always balance, like scales. Everything on one side of the equals sign must equal everything on the other side.

💬 Talk maths

There is something so satisfying about confusing an adult!

Test an adult with some missing numbers.
Secretly write a calculation nice and large, and make sure that you have the right answer, for example:

$13 - 3 \times 4 = 1$

Cover any one of the numbers with your finger, and challenge them to calculate the hidden number.

✔ Check

1. Insert the missing numbers to make these equations true.

a. $23 - \bigcirc = 15$ **b.** $\bigcirc - 7 = 11$ **c.** $6 + \bigcirc = 31$ **d.** $\bigcirc + 13 = 11$

e. $4 \times \bigcirc = 24$ **f.** $49 \div \bigcirc = 7$ **g.** $23 + 4 \times \bigcirc = 39$ **h.** $\bigcirc \div 3 - 4 = 7$

2. Solve these problems.

a. $45 = \bigcirc - 17$ **b.** $23 = 11 + 2 \times \bigcirc$ **c.** $7.3 = \bigcirc - 2.7$ **d.** $6 = \bigcirc + 9$

⚠ Problems

Brain-teaser A teacher has been collecting dinner money, but she dropped some of her own money into the bowl by accident. She knows that 25 children each gave her £1.50, and that there is £42.50 in the bowl.

Write an equation for the missing money, and use it to find out how much money the teacher should take back.

_____

Brain-buster Some children are raising money for charity. They *each* raise £5.60.
An anonymous donor says that they will match the amount raised.
The total amount raised, including the donation, is £190.40.

Write an equation for the money, and use it to find out how many children took part.

_____

Equations with two unknowns

↻ Recap

Algebra uses letters as well as numbers.
Letters are sometimes referred to as **variables**, or **unknowns**.
The letters **represent** numbers.

We can solve an equation to find the value of an **unknown** number. $16 - a = 7$

We can move letters and numbers around, but we must keep the calculation balanced.

Whatever we do to one side, we must do to the other!

How to find a:

$16 - a = 7$

$16 = 7 + a$ (we added a to each side)

$9 = a$ (we took away 7 from each side)

$a = 9$ (we wrote the equation starting with '$a = ...$')

📋 Revise

DID YOU KNOW?

In real life scientists find equations like this very, very useful.

Equations can have more than one variable or unknown.

$x + y = 6$

The problem with equations that have two variables is that there can be more than one answer.

It goes on forever!

$x = 0, y = 6$	$x = 1, y = 5$	$x = 2, y = 4$
$x = 3, y = 3$	$x = 4, y = 2$	$x = 5, y = 1$
$x = 6, y = 0$	$x = 7, y = -1$	$x = -1, y = 7$

💡 Tips

- Spend time practising balancing equations with two unknowns. It will really help you to see how they work. These equations are all the same:

$p + q = 4 \qquad p = 4 - q \qquad q = 4 - p$

Try putting $p = 3$ and $q = 1$ into each equation to check!

Talk maths

Start off by using small numbers only, but have a go with bigger ones too!

Working with a partner, choose one of the equations in the box and choose a variable each. The first person calls out a number for their letter, and the second person must find the value of the second variable.
Try it for all the equations.

$p = q + 4$

$a + b = 18$

$2x - y = 7$

$s + 3 = t - 4$

$23 - y = z$

✔ Check

1. Complete the table for each equation.

a. $y = x + 2$

x	0	1	2	3	4	5	−1	−2	−3
y									

b. $s + t = 8$

s	0	2	5	6	7	8	9	10	−1
t									

c. $p = 2q - 3$

q	0	1	1.5	2	5	10	100	−1	−10
p									

⚠ Problems

Brain-teaser Entry to the school disco is £2. The cost for disco hire is £120. The head teacher writes a formula to calculate the money they will raise.

$m = 2t - 120$

(m = the money they will make, and t = the number of tickets they will sell)

a. How many tickets must they sell to 'break even'? _____

b. How many tickets must they sell to make a profit of £50? _____

Break even means to lose nothing and gain nothing.

Brain-buster Rashid writes an equation $x^2 - y^2 = 32$. What numbers could he be thinking of if they are positive whole numbers?

What are the numbers he is thinking of? $x =$ _____ , $y =$ _____

143

Converting units

↻ Recap

Different quantities are measured in different ways.

Measure	Units of measurement	Abbreviations	
Time (years)	1 year = 12 months, = $365\frac{1}{4}$ days 1 year = approximately 52 weeks 1 week = 7 days	years = y months = m	weeks = w days = d
Time (days)	1 day = 24 hours 1 hour = 60 minutes 1 minute = 60 seconds	hours = h minutes = m	seconds = s
Length	1 kilometre = 1000 metres 1 metre = 100 centimetres 1 centimetre = 10 millimetres	kilometres = km metres = m	centimetres = cm millimetres = mm
Mass	1 kilogram = 1000 grams	kilograms = kg	grams = g
Capacity	1 litre = 100 centilitres 1 centilitre = 10 millilitres	litre = l centilitre = cl	millilitre = ml

Revise

Look at these conversion charts.

Converting length

Conversion	Operation	Example
mm to cm	÷ 10	12mm = 1.2cm
cm to m	÷ 100	256cm = 2.56m
m to km	÷ 1000	467m = 0.467km
cm to mm	× 10	3.5cm = 35mm
m to cm	× 100	1.85m = 185cm
km to m	× 1000	4.3km = 4300m

Converting mass

Conversion	Operation	Example
grams to kg	÷ 1000	250g = 0.25kg
kg to grams	× 1000	7.3kg = 7300g

Converting time

Conversion	Operation	Example
hours to days	÷ 24	48h = 2d
mins to hours	÷ 60	240m = 4h
seconds to mins	÷ 60	600s = 10m
days to hours	× 24	2d = 48h
hours to mins	× 60	7h = 420m
mins to seconds	× 60	10m = 600s

Converting capacity

Conversion	Operation	Example
cl to litres	÷ 100	7000cl = 70l
ml to litres	÷ 1000	3000ml = 3l
litres to cl	× 100	3l = 300cl
litres to ml	× 1000	2.3l = 2300ml

Tips

- You must always make sure that you are using the right units.
- To solve problems that have different quantities that can be measured, you may have to convert the units, such as 1kg + 340g = 1340g or 1.34kg

Talk maths

Work with an adult or a friend to practise converting units in your head. Using the charts on the page opposite, ask each other questions that you know will be possible to calculate mentally.

How many centimetres in 3m?

How many seconds in 5 minutes?

How many millilitres in 2.5l?

✔ Check

1. **Convert these times.**
   a. 5 hours into minutes _____
   b. 2 hours into seconds _____
   c. 510 seconds into minutes _____
   d. 1 day into seconds _____

2. **Convert these lengths.**
   a. 23m into millimetres _____
   b. 2.4km into metres _____
   c. 1km into centimetres _____
   d. 685mm into metres _____

3. **Convert these weights.**
   a. 750g into kilograms _____
   b. 32.5kg into grams _____
   c. 1g into kilograms _____
   d. 0.35kg into grams _____

4. **Convert these lengths.**
   a. 2.5l into millilitres _____
   b. 75cl into litres _____
   c. 63,425ml into litres _____
   d. 0.25l into millimetres _____

⚠ Problems

Brain-teaser The distance from Evie's front door to her school gate is exactly 242,637mm! How far is that in metres, centimetres and millimetres? (For example, 31,456mm is 31m, 45cm and 6mm.)

_____

Brain-buster How many seconds are there in a leap year? _____

Using measures

↺ Recap

Measures you should understand include:

Measure	Used for	Units
Capacity	Volumes of containers, quantities of liquid	1cl = 10ml 1l = 100cl 1l = 1000ml 1ml = 0.1cl 1ml = 0.001l
Length	Distances, lengths and areas	1km = 1000m 1m = 100cm 1cm = 10mm 1m = 0.001km 1cm = 0.01m 1mm = 0.1cm
Mass	Weights	1kg = 1000g 1g = 0.001kg
Time	Times, timetables, speed	1d = 24h 1h = 60m 1m = 60s

📋 Revise

Arranging quantities with units in powers of 10 is called **metric**. Metric systems make conversion easy by multiplying or dividing by 10, 10^2 (100) or 10^3 (1000).

Remember, you can only add like units:
1.1l + 357ml = 1.457l or 1457ml
1.45km + 257m = 1.707km or 1707m
3.23m − 122.6cm = 2.004m or 200.4cm
0.24kg + 3245g = 3.485kg or 3485g

Time is a bit different. We give answers to time in hours, minutes and seconds.
45s + 25s = 70s = 1m 10s
40m + 50m = 90m = $1\frac{1}{2}$h
3h 50m − 100m = 2h 10m

Imperial units
We sometimes use imperial units for:

lengths
1 mile = 1760 yards
1 yard = 3 feet
1 foot = 12 inches

How many kilometres is ten miles?

1 inch = 2.54cm	1cm = 0.394 inches
1 mile = 1.61km	1km = 0.621 miles

weights
1 stone = 14 pounds (lb)
1lb = 16 ounces (oz)

1lb = 0.454kg	1kg = 2.205lb
1oz = 28.35g	1g = 0.035oz

capacity
1 gallon = 8 pints

1 pint = 0.57 litres	1 litre = 1.76 pints

💡 Tips

Metric or imperial? Learn both!

- We still have imperial units in daily life. In the past, everything was measured in imperial units. These are sometimes used in other countries, but in most of the world metric units are used. The key facts in the box above are worth learning by heart so that you can do quick mental conversions.

💬 Talk maths

You will need a tape measure, some scales and a measuring jug.

Working with an adult, look for a selection of different-sized objects from around the house. Discuss whether you will measure the length, weight or capacity of each object (for some objects you can measure more than one). Write down your estimates.

✔ Check

1. Now find the actual measures of all your objects, using the appropriate equipment. With practice, you will find that your estimates become better and better.

Object	Measure	Estimate	Actual	Imperial units
Mug	Capacity	260ml	215ml	
TV	Length	82cm	93cm	
Banana	Weight	90g	130g	

Now try to calculate the imperial units for each object.

⚠ Problems

DID YOU KNOW?

1ml of water weighs 1g.
1l of water weighs 1kg.

Brain-teaser

Solve these measures problems.

If it takes 25 seconds to fill a 1 litre jug from a tap, how long will it take to fill three 250ml cups from the same tap? (You can assume that there is no time lost when changing cups.) _____

The 1 litre jug weighs 1.79kg when full of water. What is the weight of the empty jug? _____

The 250ml cup weighs 483g when full of water. What is the weight of the empty cup? _____

Brain-buster

Brian's grandad says that when he was at school he was 4 feet 11 inches tall, and weighed 6 stone, 3 pounds. Convert his height and weight to metric units.

Height _____   Weight _____

If Brian's grandad is 80 on his next birthday, calculate how many days he has lived. (There will have been 20 leap years in his life so far.) _____

How many hours has he lived? _____

How many minutes is this? _____

Perimeter and area

↺ Recap

Perimeter is the distance around the outside of a shape.
All rectangles have a width and a height.
Perimeter can be calculated with a formula:
P = 2l + 2w
Or we can say **P = 2(l + w)**.

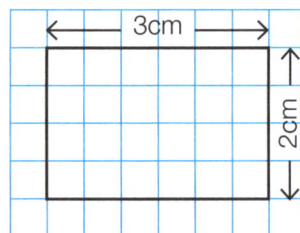

Area is measured in square units. For rectangles we multiply the length by the width: **A = lw**
For this rectangle, $P = 2(3 + 2) = 10$cm, and $A = 3 \times 2 = 6$cm.

The perimeter of a square is four times the length of a side: **P = 4s**.
The area of a square is side length times side length: **A = s²**.
The perimeter of this square is: $P = 4 \times 1.5 = 6$cm
The area of this square is: $A = 1.5 \times 1.5 = 2.25$cm²

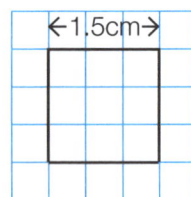

▤ Revise

Shapes that have the same perimeter do not necessarily have the same area as each other.

Remember to do any calculations in brackets first.

Shape 1

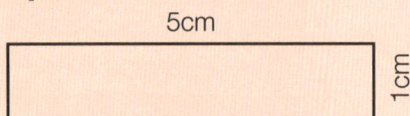

5cm

1cm

Perimeter = 2(5 + 1) = 12cm

Area = 5 × 1 = 5cm²

Shape 2

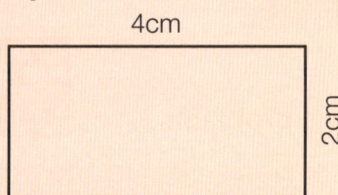

4cm

2cm

Perimeter = 2(4 + 2) = 12cm

Area = 4 × 2 = 8cm²

♡ Tips

Here's how to get your perimeters and areas right.

- Watch out for silly mistakes when you find the perimeters and areas of composite shapes.
  This shape has a square with a hole in it, joined to a rectangle.
  There are two mistakes that people often make.

  1. They include the perimeter where the shapes are joined. *Don't!*

  2. They forget to take away the area of the hole. *Do!*

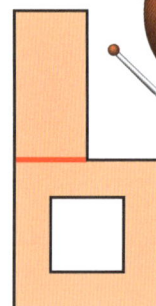

💬 Talk maths

You will need a tape measure.
Investigate the perimeter and area of different rectangles and squares around your home.
Measure their lengths and widths, then use formulae to calculate their areas and perimeters.

Object	Shape	Dimensions	Perimeter	Area
Table	Square	$s = 80$cm	320cm	6400cm²
Television	Rectangle	$l = 125$cm, $w = 75$cm	400cm	9375cm²
Door				

Explain to an adult anything you discover.

✔ Check

1. Calculate the perimeter and area of these shapes.

   a.
   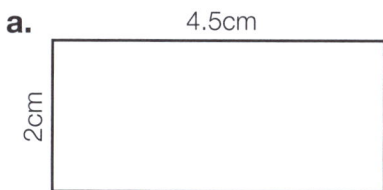
   4.5cm
   2cm

   $P =$ _____ $A =$ _____

   b.

   1.5cm

   $P =$ _____ $A =$ _____

2. Calculate the perimeter and area of these composite shapes.

   a.
   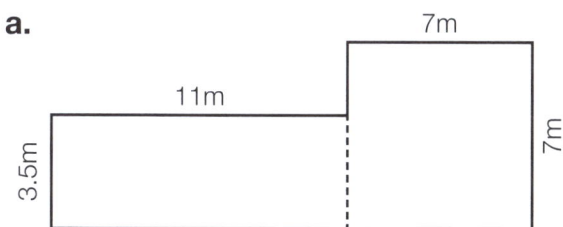
   7m
   11m
   3.5m
   7m

   $P =$ _____ $A =$ _____

   b.
   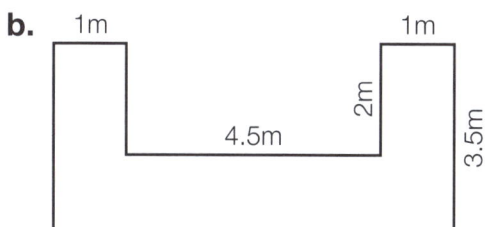
   1m   1m
   4.5m
   2m
   3.5m

   $P =$ _____ $A =$ _____

⚠ Problems

Brain-teaser Ben's rectangular garden is 5m long and has a total perimeter of 16m.

What is its area? _____

Brain-buster Some square wall tiles are 20cm wide.

How many tiles would be needed to cover a wall 3m high and 2.4m long? _____

Calculating area

↻ Recap

Area is measured in **square units**.
We can count squares for simple areas.
This rectangle has an area of 6cm².
We can use formulae for many shapes.
Formulae help us to find the areas of larger or more complex shapes.
For rectangles we multiply the length by the width: $A = lw$
The formula for the area of a square is $A = s^2$

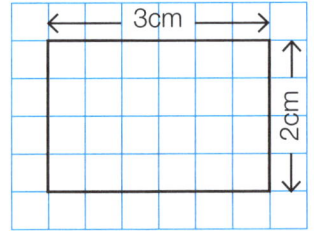

📋 Revise

The formula for the area of a triangle can be found with the formula:

$A = \frac{1}{2} bh$

b = the length of the base

h = the *perpendicular* height

Finding h can be tricky.

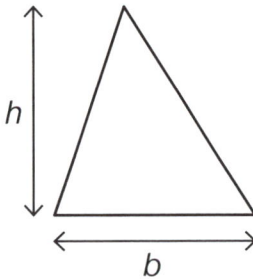

It is easier for right-angled triangles!
$A = \frac{1}{2} \times 3 \times 4 = 6cm^2$

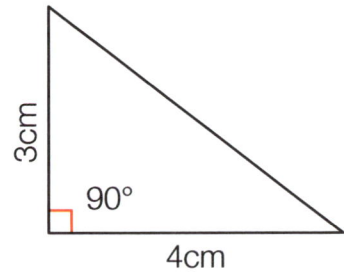

The areas of parallelograms are easy to find as long as you know the perpendicular height.

$A = hw$

Can you see why?

Imagine you had a pair of scissors and could move the dotted-line triangle.

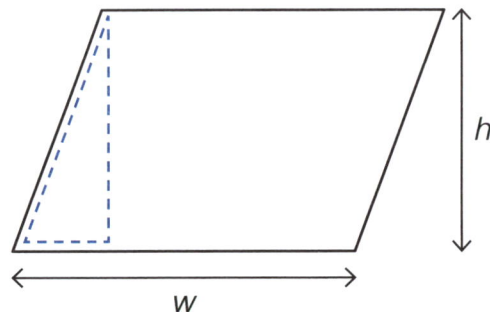

💡 Tips

Here's my tip for this area of maths...

- Think of a right-angled triangle as half of a rectangle.
It makes the formula obvious!

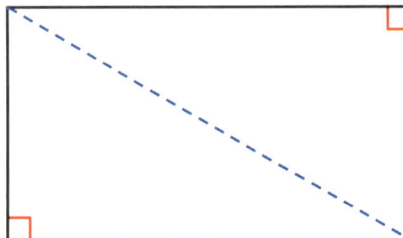

Talk maths

A rectangle is 5m long and 3.5m wide, what is its area?

The sides of a square are 2.5m. What is its area?

A parallelogram is 4.2m wide and is 3m high. What is its area?

A triangle has a 3cm base and 5cm perpendicular height. What is its area?

Use this chart to challenge a partner to mentally calculate areas. Make up your own side lengths and heights. Be sure to ask questions using the correct vocabulary.

	Rectangle	Square	Triangle	Parallelogram
Shape				
Formula	$A = lw$	$A = s^2$	$A = \frac{1}{2} bh$	$A = wh$

✔ Check

1. Calculate the areas of these shapes.

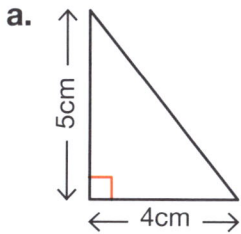

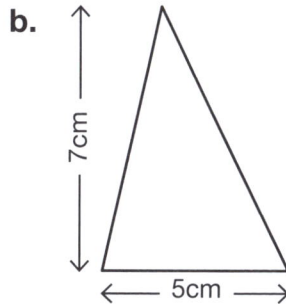

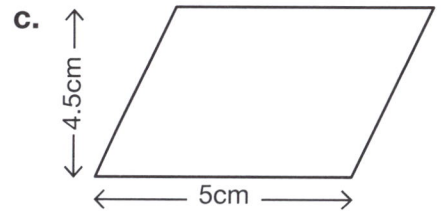

a. 5cm, 4cm

b. 7cm, 5cm

c. 4.5cm, 5cm

2. Underline which shape has the larger area.

a. A rectangle, length 7cm and width 4cm   OR   a square with sides 5.5cm

b. A parallelogram, length 9m and height 3m   OR   a triangle, base 12m and height 5m

c. A triangle, base 5cm and height 7cm   OR   a rectangle, base 6cm and height 2.5cm

⚠ Problems

Brain-teaser A carpet costs £23 per square metre.

How much would it cost to carpet a room that is 5.3m long and 4.2m wide? _____

Brain-buster In this shape, the height and base of the triangular hole are exactly half the length of the sides of the square.

What is the shaded area? _____

3 m

Calculating volume

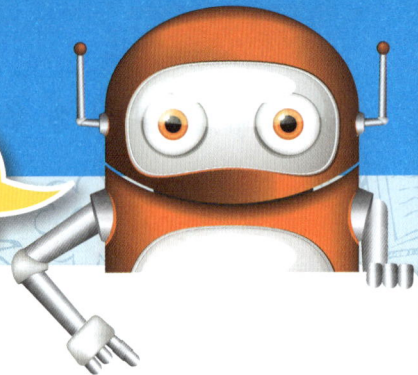

Sometimes faces are called sides.

↺ Recap

3D shapes have faces, edges and vertices.

A corner is a **vertex**. The plural is **vertices**.

Volume is the amount of space an object takes up. Volume isn't quite the same as capacity. We measure capacity in litres, centilitres or millilitres; we measure volume in cubic lengths: km³, m³, cm³, mm³.

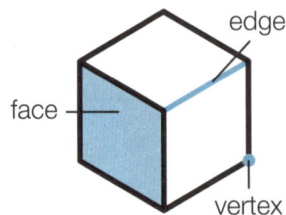

Revise

A cubic centimetre is a cube that has length, width and height all equal to 1cm.

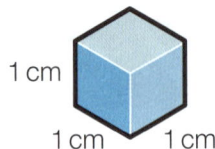

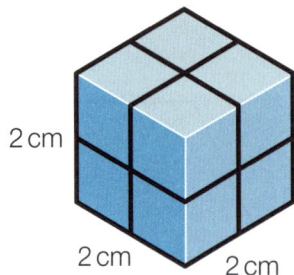

A cube that has sides of length 2cm has a volume of 8cm³. A *cube* is a 3D shape that has all sides the same length.

You need to be careful with units.
$1cm^3 = 10mm \times 10mm \times 10mm = 1000mm^3$
$1m^3 = 100cm \times 100cm \times 100cm = 1{,}000{,}000cm^3$

We can use formulae for calculating the volumes of cubes and cuboids.

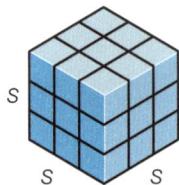

Volume of cube = s^3 (s = length of one side)
A cubes of with side 3cm has a volume of
$3cm \times 3cm \times 3cm = 27cm^3$

Volume of cuboid = whl (w = width, h = height, l = length)
A cuboid with width 4cm, height 2cm and length 5cm has a volume of $4cm \times 2cm \times 5cm = 40cm^3.$

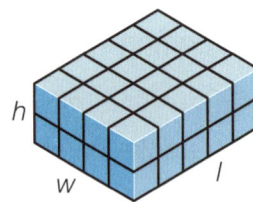

A *cuboid* has rectangular faces, but they are not all the same size.

💡 Tips

Here's my advice for perfecting your 3D drawings...

- Drawing shapes to look 3D is called *isometric drawing*. The trick is to draw one end face, and then draw the edges as parallel lines.

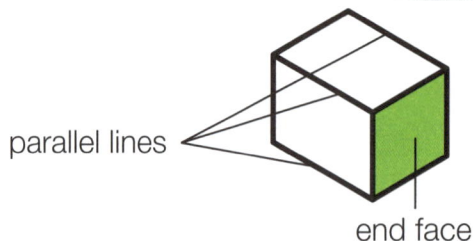

parallel lines

end face

Talk maths

Work with an adult or a friend to discuss how you might estimate the volume of large objects. For example in a bathroom you could estimate that the room is 3m high, 4m long and 2m wide. So the volume of the bathroom could be estimated as 24m³.

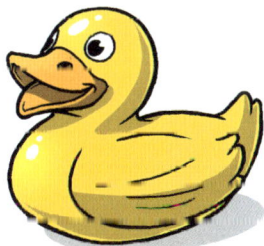

What about the volume of a bath?

Or even the volume of your house?

✔ Check

1. Use a pencil and ruler to draw each of these shapes.

   a. A cube with side length 3cm

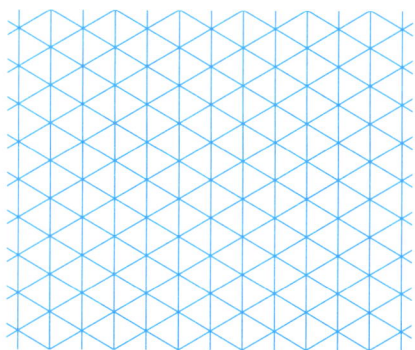

   b. A cuboid, length 5cm, height 2cm, width 4cm

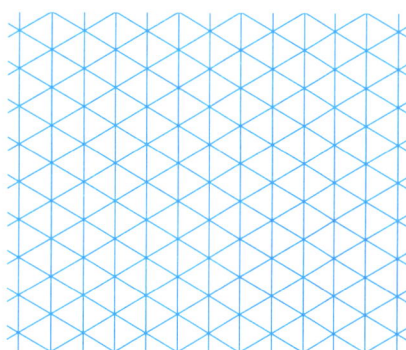

2. Calculate the volume of these shapes.

   a. Cube, side 6cm _____
   b. Cuboid, l = 6m, w = 4m, h = 1.5m _____
   c. Cube, side 10m _____
   d. Cuboid, l = 9cm, w = 5cm, h = 2cm _____
   e. Cube, side 12mm _____
   f. Cuboid, l = 60mm, w = 30mm, h = 5mm _____

3. How many cubic millimetres are there in 1m³? _____

⚠ Problems

Brain-teaser A cube-shaped packing crate is 0.5m long on each side.

Calculate its volume: in m³ _____   in cm³ _____

Brain-buster A wooden cuboid has a square-shaped hole cut right through its middle.

What is the volume of the remaining wood? _____

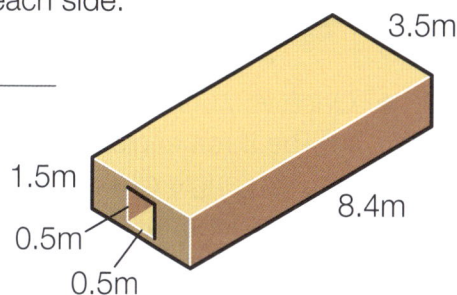

3.5m

1.5m

0.5m

0.5m

8.4m

Angles

↻ Recap

We measure angles with a protractor.

A right angle is 90°. A straight line is 180°.	Acute angles are between 0° and 90°. Obtuse angles are between 90° and 180°.
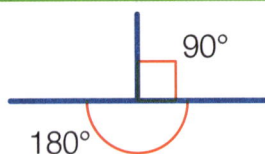 90° 180°	 obtuse 130° acute 50°

Angles greater than 180° are called *reflex* angles.	A complete turn is 360°.
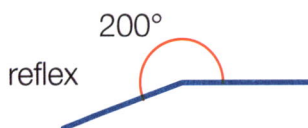 200° reflex	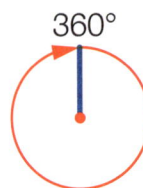 360°

📄 Revise

Angles that form a right angle add up to 90°.	Angles on a straight line add up to 180°.
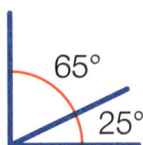 65° 25°	 118° 62°
Vertically opposite angles are equal.	Similar angles on parallel lines are equal.
 30° 30°	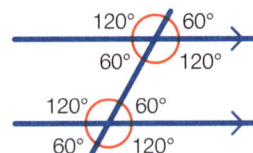 120° 60° 60° 120° 120° 60° 60° 120°

💡 Tips

- Once you understand how angles work, identifying and constructing shapes is easy!
  - The three angles of a triangle add up to 180°.
  - Each angle of an equivalent triangle = 60°.
  - The four angles of a quadrilateral add up to 360°.
  - Each angle of square and rectangle = 90°

When drawing shapes, let's talk angles.

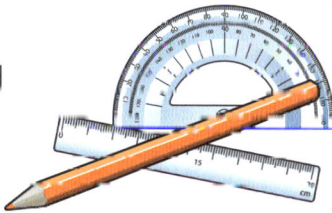

Try out your presentation on an adult!

Talk maths

You will need a paper, a pencil, a ruler and a protractor.
Prepare a presentation that will explain different types of angles from page 154.

✔ Check

1. **Use a protractor to draw these angles, and then name them.**

a. 90°

b. 23°

c. 167°

Name: _____

Name: _____

Name: _____

2. **Write down the value of each angle marked with a letter.**

a.

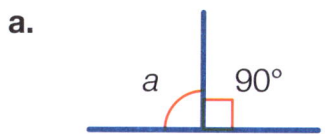

b.

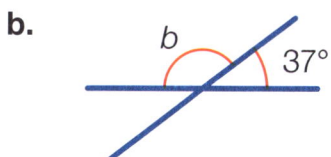

c.

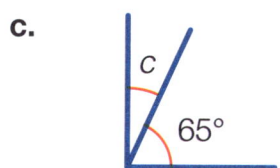

angle a = _____

angle b = _____

angle c = _____

⚠ Problems

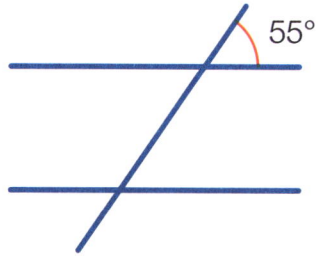

55°

Brain-teaser Two parallel lines are intersected by another line. There are eight different angles.
Without using a protractor, complete the size of every angle in the diagram.

Brain-buster This shape is a parallelogram – its opposite sides are parallel. How can you use it to prove that the four angles of a quadrilateral add up to 360°?

_____

_____

Properties of 2D shapes

↺ Recap

There are different types of triangles. Each has different properties.

Equilateral	Isosceles	Right-angled	Scalene
All sides equal All angles 60°	Two sides equal Two angles equal	One angle equals 90°	All sides different All angles different

Quadrilaterals also have different properties.

Square	Rectangle	Rhombus	Parallelogram	Kite	Trapezium
All sides equal All angles 90°	Opposite sides equal All angles 90°	All sides equal Opposite angles equal	Opposite sides equal and parallel Opposite angles equal	Adjacent sides equal	Only one pair of parallel sides

📄 Revise

Internal angles is a posh name for angles at the corners.

We say that different 2D polygons have different properties.
The sum of internal angles is the same for each shape, whether irregular or regular.

Triangle	Quadrilateral	Pentagon	Hexagon	Heptagon	Octagon
3 sides	4 sides	5 sides	6 sides	7 sides	8 sides
Angles add to 180°	Angles add to 360°	Angles add to 540°	Angles add to 720°	Angles add to 900°	Angles add to 1080°

💡 Tips

Think triangles!

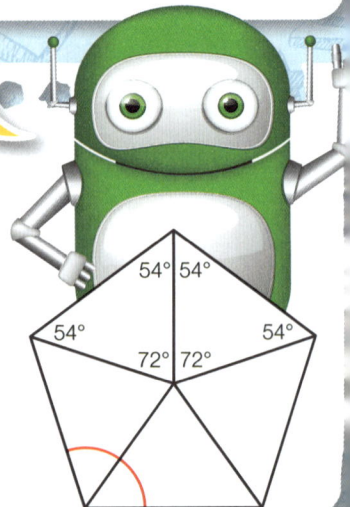

- Take any regular shape and divide it into equal triangles.

  The total of the angles at the centre must be 360°, so we can work out each angle around the centre by dividing 360° by the number of triangles. The angles of a triangle all add to 180°, so we can work out the other angles of the triangle, and then the angles at each corner of the shape.

  Look at this regular pentagon. Can you see why each internal angle is 108°?

54° 54°
54° 54°
72° 72°

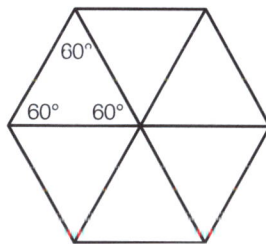

 Talk maths

You will need a protractor, a ruler, a pencil and paper.
Work with a friend or an adult to investigate the angles
inside the six regular shapes.
Read though the information and tips on the previous page,
and discuss how you will approach your investigation.

6 × 60° = 360°
Internal angle:
60° + 60° = 120°

✔ Check

1. **What is the difference between a regular and an irregular polygon?**

_____

2. **Label these polygons, and say if each is regular or irregular.**

a.

b.

c.

_____   _____   _____

d.

e.

f.

_____   _____   _____

⚠ Problems

Brain-teaser How can you prove that a square is made of four identical right-angled triangles?

_____

_____

Brain-buster Jade says that a regular hexagon is made of six equilateral triangles.
Explain whether she is right or wrong, and why.

_____

_____

Drawing 2D shapes

↻ Recap

A polygon is any straight-sided 2D shape. These are regular polygons. For each shape the internal angles are the same size and the sides are the same length.

Triangle	Quadrilateral	Pentagon	Hexagon	Heptagon	Octagon
3 sides	4 sides	5 sides	6 sides	7 sides	8 sides
Angles add to 180°	Angles add to 360°	Angles add to 540°	Angles add to 720°	Angles add to 900°	Angles add to 1080°

📄 Revise

To draw any triangle you need to know two angle sizes and one side length. Or two side lengths and one angle.

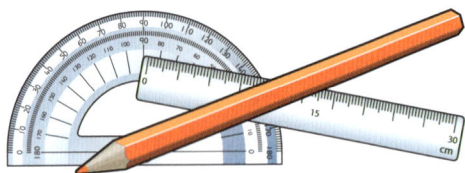

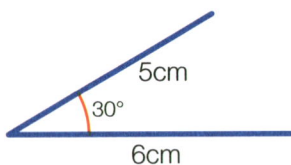

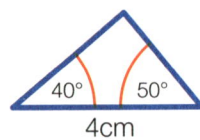

5cm 30° 6cm

40° 50° 4cm

To draw a square, rectangle, rhombus or parallelogram you only need to know one angle size and two side lengths.

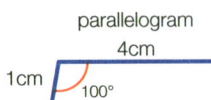

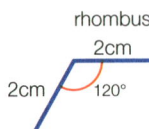

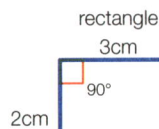

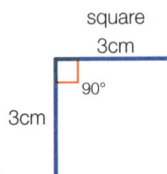

parallelogram 4cm 1cm 100°

rhombus 2cm 2cm 120°

rectangle 3cm 90° 2cm

square 3cm 90° 3cm

There is a link between geometry and algebra, because we can write formulae for different shapes. If the angles of a triangle are a, b and c, we can say $a + b + c = 180°$.

> Can you think of formulae for the angles in other regular polygons?

💡 Tips

- You need to know how to draw 2D shapes. Remember that all regular pentagons (five sides), hexagons (six sides), heptagons (seven sides) and octagons (eight sides) are all made of identical triangles.
- Also remember that all the angles at the centre add up to 360°.

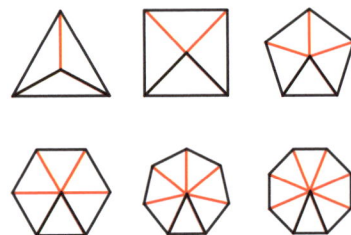

Talk maths

Play this game with a partner. You will need pencils, paper, a ruler and a protractor. Take turns to challenge each other to construct shapes, giving verbal instructions. Remember to give enough information, for example: *Draw an isosceles triangle with a base of 6cm and two angles of 65°.*

> Draw a rhombus with sides of 5cm and two angles of 125°.

> Draw a regular hexagon with sides of 4cm.

✔ Check

1. Draw an equilateral triangle with each side 4cm.

2. Draw a rhombus, with sides 3cm and the larger angle = 120°.

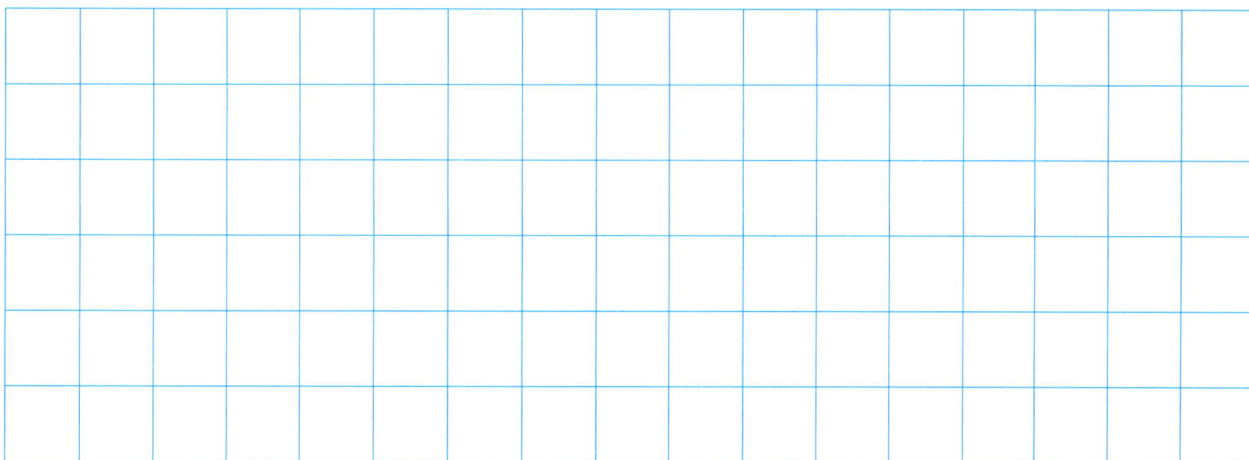

3. Explain how you would construct a regular octagon.

_____

_____

⚠ Problems

Brain-teaser The five internal angles of a regular pentagon add up to 540°. A ten-sided shape is called a decagon. What will the internal angles of a regular decagon add up to? Show your working out.

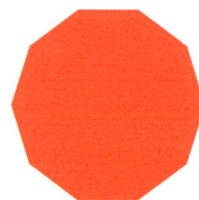

_____

_____

Brain-buster Can you write a formula for calculating the size of each angle in a regular polygon, where a = the angle and n = the number of sides? _____

3D shapes

↻ Recap

3D shapes have different properties which identify them.

Shape							
Name	Cube	Cuboid	Cone	Sphere	Cylinder	Triangular prism	Square-based pyramid
Faces	6	6	2	1	3	5	5
Edges	12	12	1	0	2	9	8
Vertices	8	8	0	0	0	6	5

▤ Revise

Some 3D shapes can be represented by **nets**. A net is a 2D drawing of the shape as if it has been taken apart, or unfolded. The skill is in thinking about which edges meet.

There is more than one way to make a net, and plenty of ways to get it wrong! Look at these cube nets.

You cannot make accurate nets for spheres or cones because they have curved faces. Try peeling an orange and laying it flat – it cannot be done accurately.

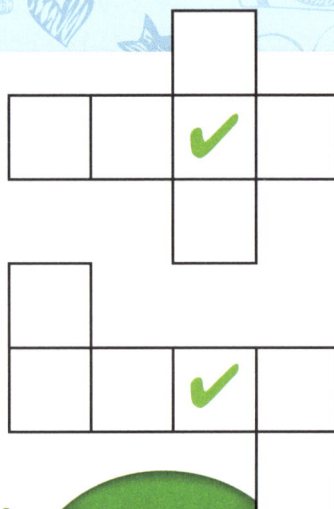

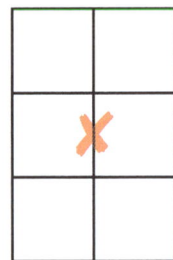

That is why maps of the world are tricky to make.

Mark it. Tab it. Net it!

💡 Tips

- When looking at, or drawing, nets, use marks to help you see if the sides match up correctly. Use tabs to help you to join the faces together.

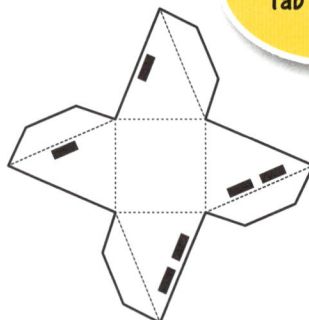

Talk maths

You will need squared paper, a ruler and a pencil.
Work with a partner to construct three cuboids of different sizes, making nets for each one. When you have finished, discuss the steps you took to make a successful net, then explain these instructions to someone else and see if they can make a net using your advice.

> Remember, you only need one tab to join two faces.

✔ Check

1. Add tabs to these nets so that the faces would join together.

   a. Pyramid

   b. Prism

   c. Cuboid

2. Draw a net for a cube with 2cm edges. Include tabs to join the faces together.

⚠ Problems

Brain-teaser Write instructions for how to make a paper model of a square-based pyramid.

_____

_____

_____

Brain-buster Ryan has a sheet of paper 30cm long and 20cm wide.

a. What is the largest cuboid he can make from it? _____

b. How much paper will be wasted (in cm²)? _____

c. What will be the volume of the cuboid? _____

Circles

↻ Recap

A circle is a single line that is always the same distance from its centre.

We can draw circles using a pair of compasses.

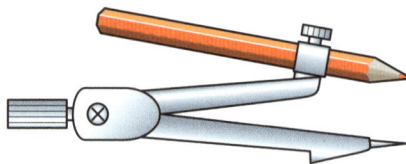

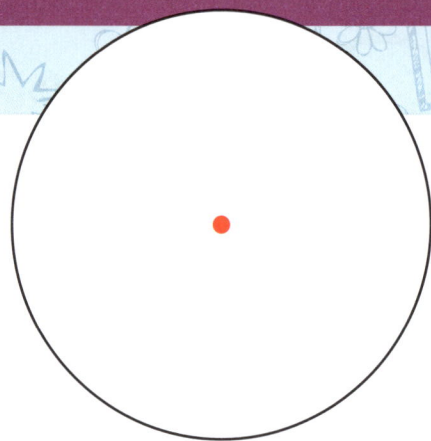

📄 Revise

The edge of a circle is called the **circumference**.

The distance across the centre of a circle is called the **diameter**.

The distance from the circumference to the centre is called the **radius**.

The diameter is twice the length of the radius.

We write this using the formula: $d = 2r$

You can estimate the circumference of a circle using thread or string, and you can estimate area by counting squares and parts of squares.

diameter

radius

circumference

Remember, circles are 2D shapes, and spheres are 3D.

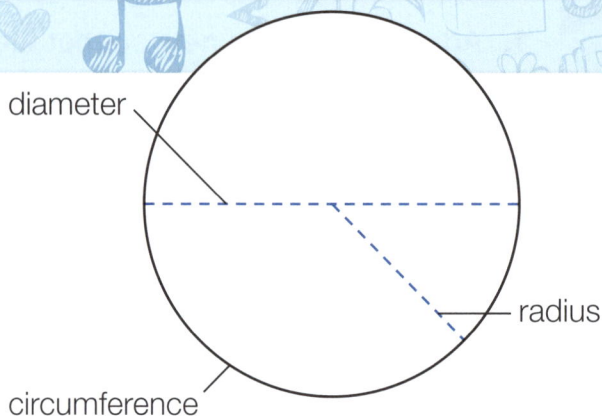

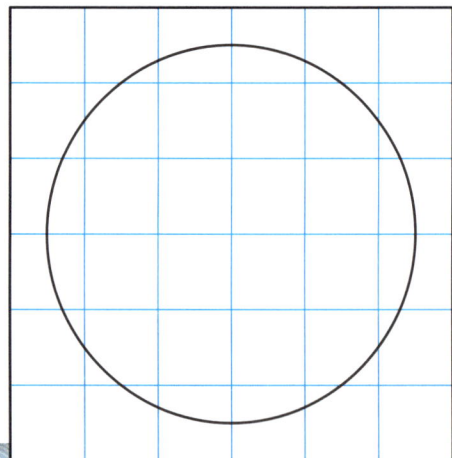

💡 Tips

- You can draw a circle using only string and a pencil. Tie the pencil to the string and hold the string tightly where you want the centre of the circle to be.

  Try drawing different-sized circles just using string.

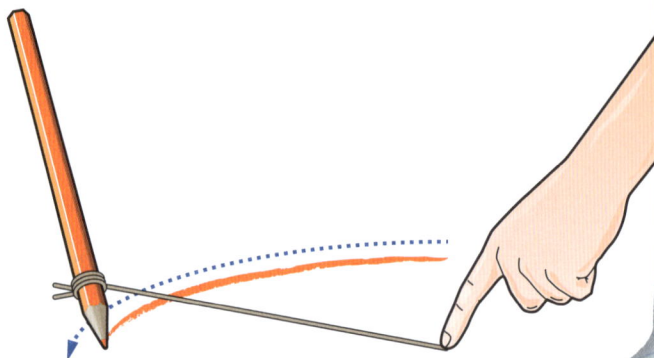

The string might help you measure the circumference.

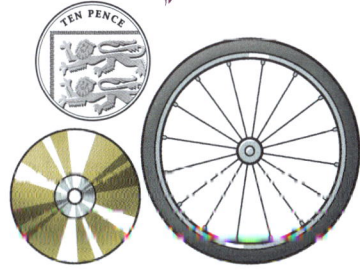

Talk maths

You will need some string, a ruler, a compass and a pencil.

Working with a partner, find a collection of approximately ten circular objects. Using your equipment, find the radius, diameter and circumference for each one. Make sure you agree on each measurement before you add it to a table.

Object	Radius	Diameter	Circumference
10p	1.25cm	2.5cm	7.85cm
DVD	6cm	12cm	37.5cm
Bike wheel	25cm	50cm	157cm

Discuss the connection between the size of the circumference and the size of the diameter or the radius?

✔ Check

1. Explain these terms.

   a. radius: _____

   b. diameter: _____

   c. circumference: _____

2. If a circle has a radius of 3.5m, what is its diameter? _____

3. A circular field has a diameter of 1.5km. What is its radius? _____

⚠ Problems

Brain-teaser Aaron says that the circumference of any circle is just over five times its diameter. Looking at this circle, would you say he is right? Explain your answer.

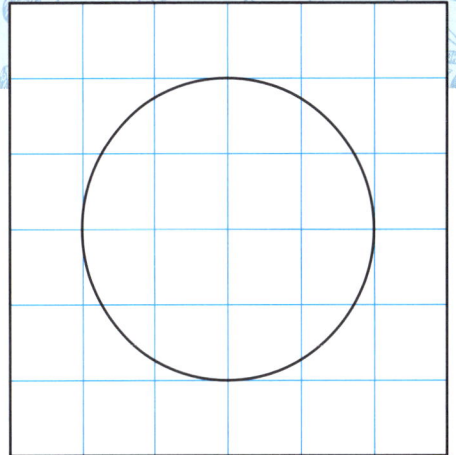

_____

_____

Brain-buster Meena says that the area of any circle is approximately three times the radius squared, or $3r^2$. Looking at this circle, would you say she is right? Explain your answer, using calculations if necessary.

_____

_____

Positive and negative coordinates

Remember: points on a grid are always shown with the x-coordinate first, and then the y-coordinate.

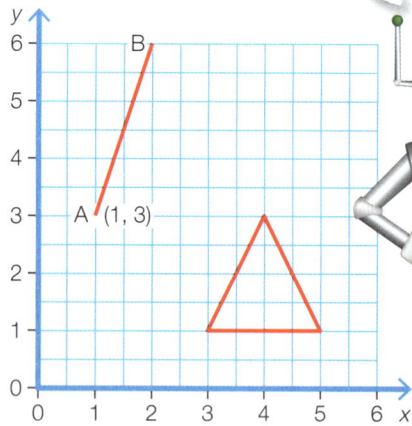

↻ Recap

We can plot points anywhere on a coordinate grid to make lines or to show the vertices of shapes.

The co-ordinates of B are (2,6).

The triangle's vertices coordinates are (3, 1), (5, 1) and (4, 3).

📄 Revise

The axes are like thermometers!

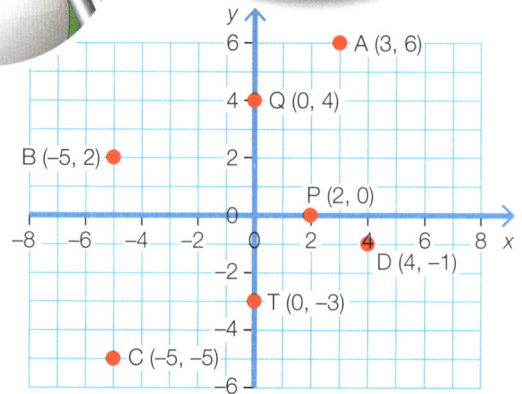

Grids can have negative axes too. They are just like number lines.

We say the coordinate grid has four **quadrants**.

Coordinates are positive and negative according to which quadrant they are in.

Remember, the point where the axes meet is called the origin. The coordinates of the origin are (0, 0).

Look at the points on the coordinate grids. Each one has its coordinates next to it.
The coordinates of A are (3, 6).

💡 Tips

- Each quadrant will always be positive or negative for x and y.
  1st quadrant: x and y positive
  2nd quadrant: x negative, y positive
  3rd quadrant: x and y negative
  4th quadrant: x positive, y negative
- Remember, for reading coordinates it's along first, then up.

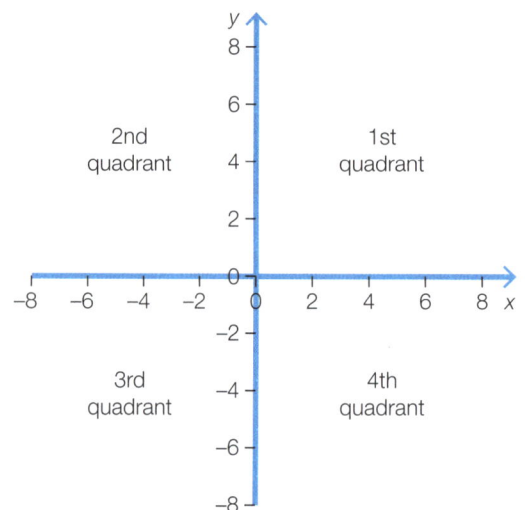

Talk maths

Where is the point (−5, −8)?

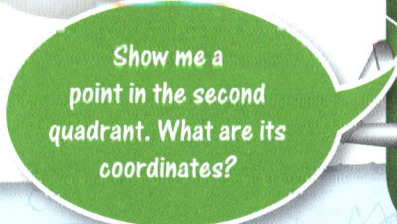

Draw a coordinate grid with four quadrants, with each axes going from −8 to +8, or use one of the grids on these pages. Working with a partner, challenge each other to identify points in particular quadrants.

Show me a point in the second quadrant. What are its coordinates?

✔ Check

1. **a.** Write the coordinates of each point marked on the coordinate grid.

   A: (_____ , _____) B: (_____ , _____)

   C: (_____ , _____)   D: (_____ , _____)

   b. What shape do they make?

   _____

   c. Write the coordinates of the centre of the shape.

   _____

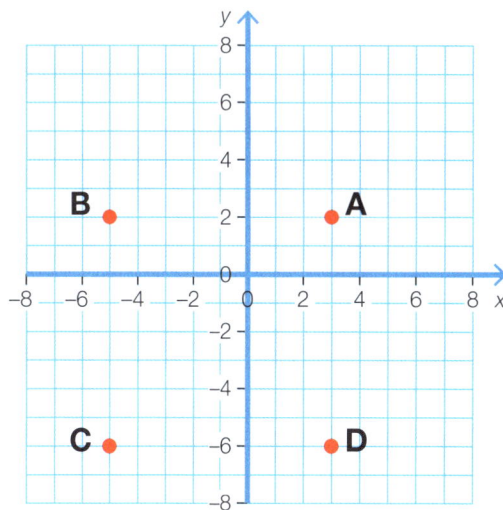

2. **a.** Mark these points on the grid.

   P (3, 5) Q (−1, 5) R (−4, −1) S (0, −1)

   b. What shape do they make? _____

⚠ Problems

Brain-teaser What shape do these points make when joined together?

A (0, 6), B (−3, 4), C (0, −2), D (3, 4) _____

Brain-buster A rectangle's centre is at the point (2, 1) and the coordinates of one vertex is at (7, 6).
Write the coordinates of the other three vertices and say which quadrant each is in.

	Quadrant	Coordinates
Vertex 1	1	(7, 6)
Vertex 2		
Vertex 3		
Vertex 4		

Reflecting and translating shapes

↺ Recap

Why did only the y-coordinate change?

When we translate points, we say how much the x and y coordinates change.
For example, A (2, 1) to A^1 (8, 4)
x has increased by +6
and y has increased by +3.

When we reflect points, the line of reflection acts like a mirror, and the coordinates of the reflected points change. For example, B (3, 6) to B^1 (3, 8)

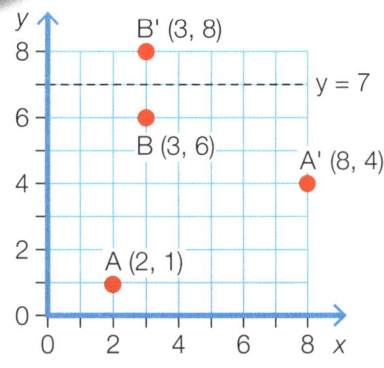

📄 Revise

For four-quadrant grids, we can translate and reflect in the same way.

Rectangle PQRS has been **translated**. Each vertex of the rectangle moves by the same amount. Can you see what the missing numbers are?
For P^{1}Q^{1}R^{1}S^1, $x = +9$ $y = -10$

We can also reflect points and shapes in the x-axis and y-axis.
Notice how the triangle ABC has changed; it has been **reflected** in the x-axis.
WXYZ has been **reflected** in the y-axis. What has happened to each vertex of the square?

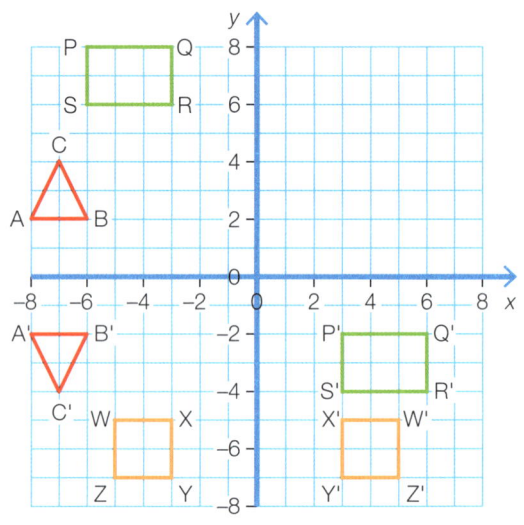

💡 Tips

- **Reflections**
  For reflections in the y-axis, only the x-coordinates change: they reverse their sign.
  For reflections in the x-axis, only the y-coordinates change: they reverse their sign.
- **Translations**
  We can write translations as, for example, $x - 8, y + 6$, or whatever the translation is.
  Remember, for shapes, every vertex will be translated by the same amount.

Talk maths

You will need a pencil, a ruler and a rubber. Carefully draw a shape on this grid and challenge someone to reflect or translate it by giving them precise instructions.

Ask them to give you the new coordinates of the shape.

Draw your shapes gently and then rub them out so that you can repeat the challenge a few times. As an extra tricky challenge, reflect a shape and then translate it too.

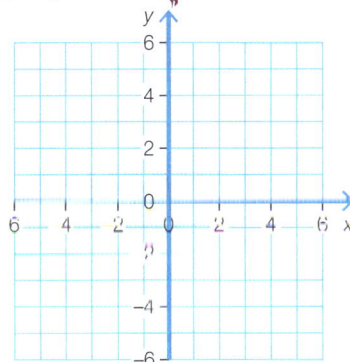

Reflect the square in the y-axis.

Translate the triangle by $x + 3, y - 2$.

✔ Check

1. Using the coordinate grid opposite.

   a. Translate the shape PQRS by $x - 3$ and $y + 2$.

   b. Reflect the shape WXYZ in the y-axis.

   c. Reflect the shape WXYZ in the x-axis.

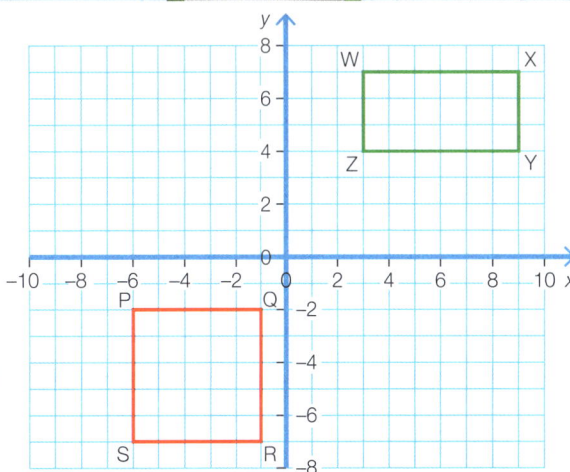

2. A triangle A (6, 2) B (0, 5) C (−1, −3) is reflected in the x-axis to create triangle $A^1B^1C^1$. Write the coordinates of each new vertex.

   A^1 (_____ , _____) B^1 (_____ , _____) C^1 (_____ , _____)

3. The triangle D (0, 0) E (3, −3) F (−1, −2) is translated by $x - 2, y - 4$ to create triangle $D^1E^1F^1$. Write the coordinates of each new vertex.

   D^1 (_____ , _____) E^1 (_____ , _____) F^1 (_____ , _____)

⚠ Problems

Brain-teaser What is unusual about reflecting the square P (3, −2) Q (3, −8) R (−3, −8) S (−3, −2) in the y-axis?

_____

Brain-buster Sam says that reflecting the square ABCD in the x-axis and then in the y-axis is the same as translating it $x + 9, y + 7$. Is he right?

Explain your answer. _____

_____

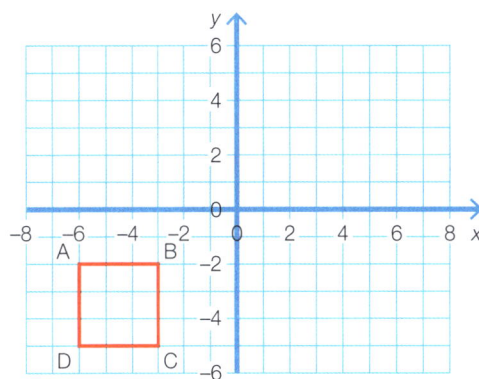

Pie charts

Different graphs are used for different situations and different types of data.

↻ Recap

We can represent information and data in different types of charts and graphs.

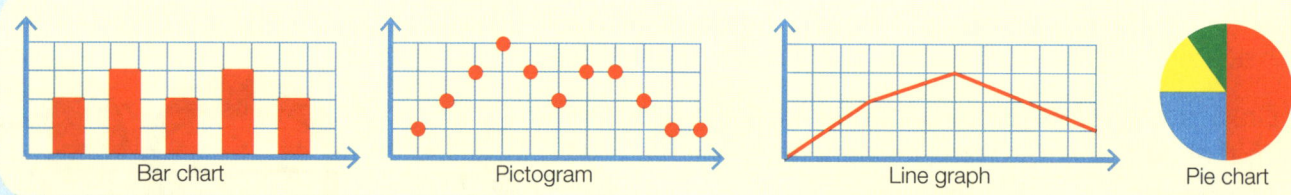

Bar chart   Pictogram   Line graph   Pie chart

📄 Revise

Remember – a complete rotation has 360°.

Pie charts use fractions of circles to represent quantities. They are great for helping us to see proportions at a glance.

This pie chart shows the different proportion of journeys to school made by all children in Britain.

Because a complete rotation is 360°, any fraction or percentage is shown as an angle, as a part of 360°.

Look at the same information in a chart.

Transport	Walking	Car	Bike	Bus
Fraction	$\frac{1}{2}$	$\frac{1}{4}$	$\frac{1}{8}$	$\frac{1}{8}$
Percentage	50%	25%	12.5%	12.5%
Angle on pie chart	180°	90°	45°	45°

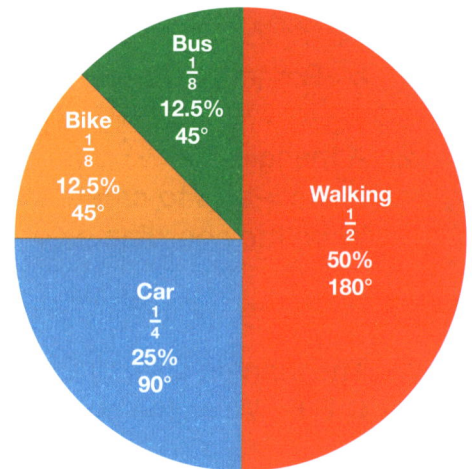

Although this pie chart doesn't show us the actual number of journeys, we can still work these out. If the total number of journeys to school each day in Britain was 10 million, we can use the pie chart to calculate numbers for each of the journey types. For example, 5 million children must walk, because 5 million is half of 10 million.

💡 Tips

- To understand pie charts you need to convert angles, fractions and percentages. Try to learn the main ones.

Angle	3.6°	18°	36°	45°	90°	180°	360°
Fraction	$\frac{1}{100}$	$\frac{1}{20}$	$\frac{1}{10}$	$\frac{1}{8}$	$\frac{1}{4}$	$\frac{1}{2}$	$\frac{1}{1}$
Percentage	1%	5%	10%	12.5%	25%	50%	100%

Talk maths

Work with a partner, challenging each other to convert percentages and fractions into angles on a pie chart. Then try converting angles on a pie chart into fractions and percentages.

What angle would 55% be on a pie chart?

What angle would five twelfths be on a pie chart?

What percentage does 60° on a pie chart represent?

What fraction would 200° on a pie chart be equivalent to?

✔ Check

1. 48 children took part in a survey on favourite animals. The results are shown on the pie chart. Complete the table to show how many children prefer each type of animal.

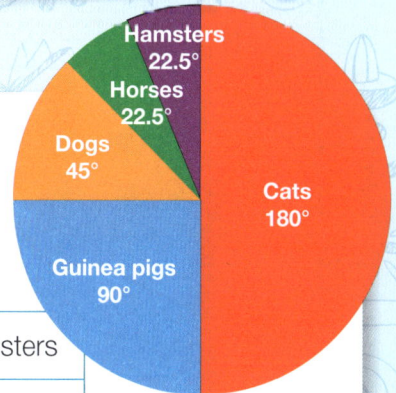

Hamsters 22.5°
Horses 22.5°
Dogs 45°
Cats 180°
Guinea pigs 90°

Animal	Cats	Guinea pigs	Dogs	Horses	Hamsters
Angle	180°	90°	45°	22.5°	22.5°
Children					

2. You will need a protractor to do this activity.
A family of five are on holiday. They need to catch a taxi and they only have loose change in their pockets. Draw a pie chart to show the proportion each person contributes to the total amount.

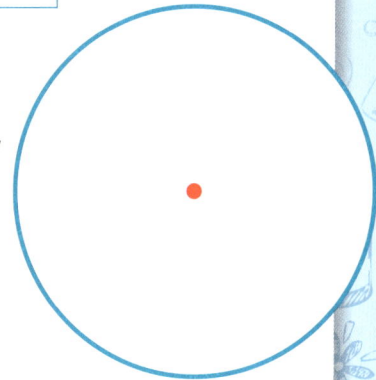

Mum	Dad	Paul	Lizzie	Mary
£1.80	5p	45p	£1.20	10p

⚠ Problems

The pie chart shows the population of the world by continent.

Brain-teaser Without measuring, can you estimate the angle for Europe in the pie chart?

_____

Brain-buster If the population of the world is 7 billion, estimate the population of each continent, to the nearest tenth of a billion.

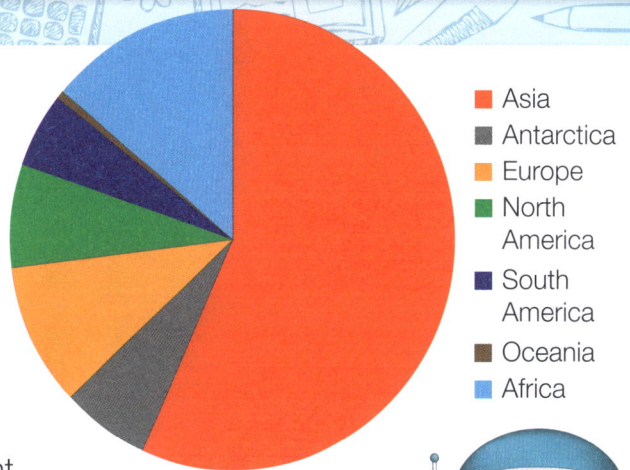

- Asia
- Antarctica
- Europe
- North America
- South America
- Oceania
- Africa

Asia	Africa	Europe	Oceania	North America	South America
4.2 billion					

Your estimates should add up to around 7 billion!

169

Line graphs

↺ Recap

Line graph for a bike ride

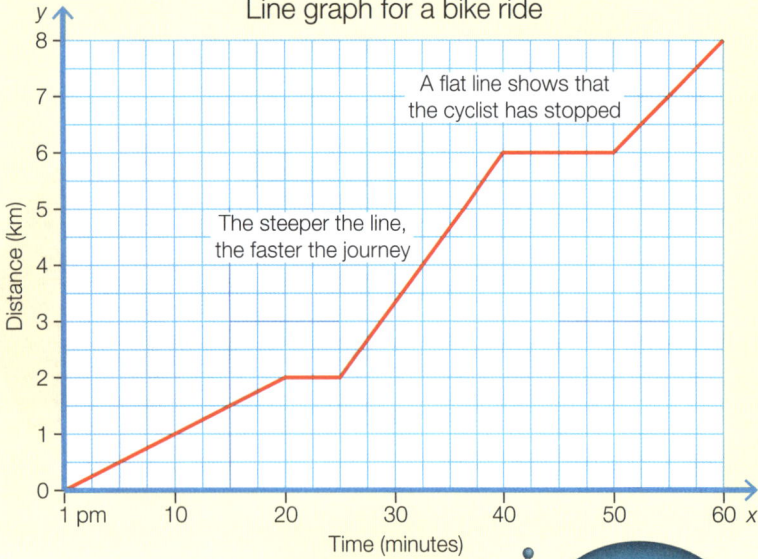

The steeper the line, the faster the journey

A flat line shows that the cyclist has stopped

Line graphs are useful for showing how things change over time, such as temperature, growth and speed. Normally time is shown along the horizontal x-axis.

Remember, to read a graph, you go along the x-axis and up the y-axis.

This graph shows the time taken for an 8km cycle ride.

Find these bits of information on the graph.

- The journey starts at 1pm.
- After 20 minutes the cyclist stops for five minutes.
- The cyclist travels fastest from 25 minutes to 40 minutes.
- The cyclist stops again after 40 minutes.
- The journey finishes at 8km.

Look carefully at the scale on each axis.

📄 Revise

Line graphs can also be used for converting similar quantities that have different units, such as temperatures, distances and currencies.

This graph can be used for converting US dollars to pounds.

The direction of the line is affected by how many dollars there are for every pound. In this graph

£1 = $1.5   £5 = $7.5 $9 = £6

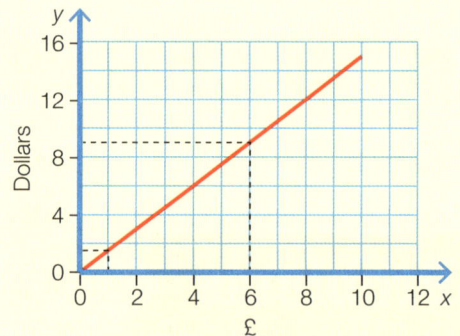

How many dollars would you get for £5?

How many pounds would you get for $9?

💡 Tips

- Use a pencil and a ruler for accurate conversions. Remember that you need to read graphs carefully – use a ruler to help you read horizontal and vertical coordinates.

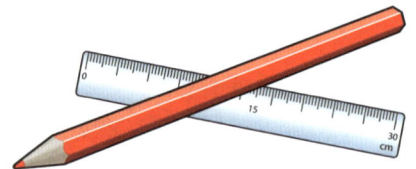

Talk maths

How many kilometres in 50 miles?

Practise converting miles to kilometres, and kilometres to miles, using this line graph.

How many kilometres are there in 2 miles? How many miles are there in 5 kilometres? Can you use the graph to work out larger distances?

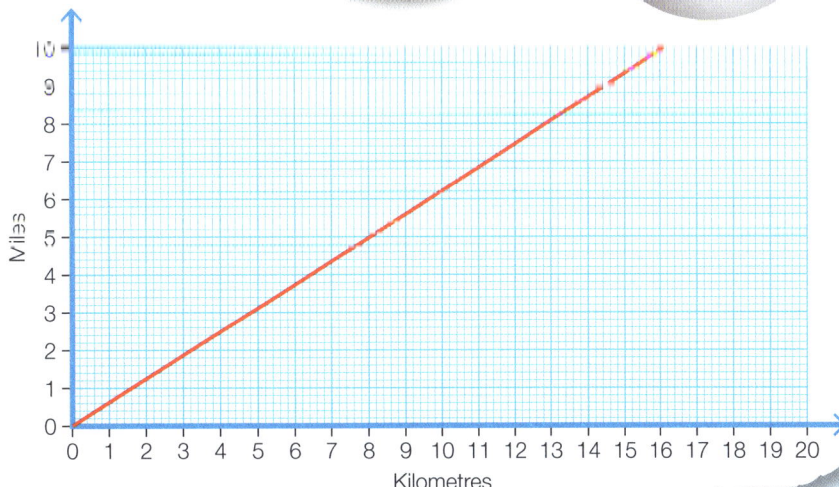

✔ Check

1. **The line graph shows the distance travelled on a charity walk.**

a. Why is the line horizontal from 11am to 11.30am, and from 2pm to 3pm?

_____

_____

b. How far did the walkers travel from 3pm to 5pm? _____

c. The fastest part of the journey was from 9am to 11am. At what speed were the walkers travelling then? _____

d. Another group of walkers started walking at 12 noon and walked at a constant speed and arrived at the end exactly the same time as the other group. Draw their journey on the graph with a dotted line.

e. What was the speed of the second set of walkers? _____

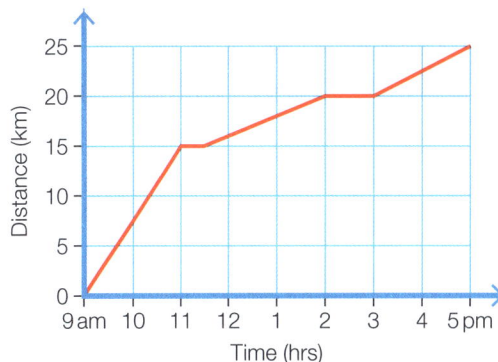

⚠ Problems

This is the formula for converting temperatures from Fahrenheit to Celsius: $C = \frac{5}{9} \times (F - 32)$
It can be used to create a line graph.
Draw the line graph on square paper and then use it to solve these problems.

Brain-teaser Water boils at 100°C. What is that in degrees Fahrenheit? _____

Brain-buster Where does the line cross the *y*-axis? _____

Where does the line cross the *x*-axis? _____

Averages

⟳ Recap

We can collect data and represent it in tables, charts and graphs.

For example, this bar chart shows the number of vegetarian school lunches eaten each day for a week.
We call this collection of information a data set.

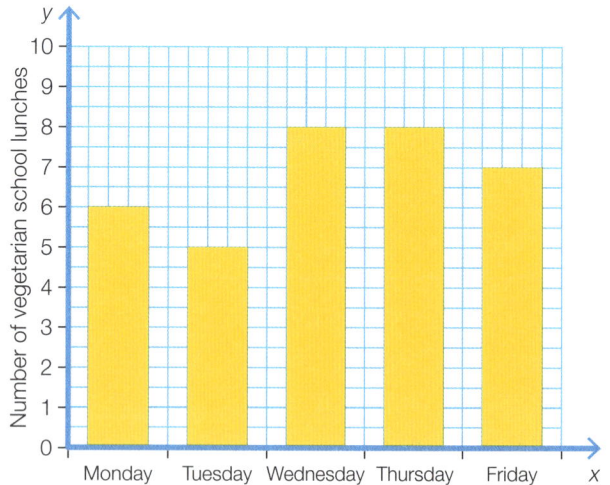

📋 Revise

A mean is the average of the data set.
To find the mean, add together all of the numbers and then divide it by how many numbers there are.

Day	Monday	Tuesday	Wednesday	Thursday	Friday
Vegetarian lunches	6	5	8	8	7

Using the above definitions, we can find out the mean for this data set.
Mean = $(6 + 5 + 8 + 8 + 7) ÷ 5$
$= 34 ÷ 5 = 6.8$

We can say, on average, 6.8 vegetarian lunches are eaten each day.

💡 Tips

- Remember, the mean average is not always a whole number.

  Mean averages are useful for comparing things. For example, the number of people going on holiday in the summer, is higher than at other times. The mean average for holidays in a year would be very different to just the summer months.

Talk maths

7 8 9 9
7 10 11 8 10

With a partner, discuss what the mean of this set of data will be.

Practise with different data sets. Roll a dice four or five times to generate a new set of numbers.

Can you find all of these without using a pencil and paper?

✔ Check

1. Seven children were asked how many pieces of fruit they eat each week. The results are shown here. Find the mean using the data. Show your working out.

14 4 12
12 6 0 8

_____

2. A park-keeper counts the number of flowers in each flowerbed.

Flowerbed 1	Flowerbed 2	Flowerbed 3	Flowerbed 4	Flowerbed 5	Flowerbed 6
23	25	20	23	26	28

   a. Find the total number of flowers in the park. _____

   b. Find the mean number of flowers per flowerbed. _____

⚠ Problems

Brain-teaser Just before the summer holidays, ten Year 6 children each estimate (to the nearest five) how many books they have read in their time in the Juniors. Calculate the mean.

Aaron	Fahad	Beth	Jin	Eva	Scarlett	Mason	Sam	Jayden	Zac
45	50	75	35	50	90	40	50	45	80

_____

Brain-buster Gemma is reading a novel and wants to estimate how many words she has read. She counts the words on six different pages: 274 286 259 262 294 272

What is the average number of words per page? _____

If the book is 386 pages long, and 20 of the pages are only half full (because they start or end a chapter), estimate how many words in total are in the book. _____

English glossary

A

active voice is when the subject does the action. *The school arranged a visit.*

adjectives are sometimes called 'describing words' because they pick out features of nouns such as size or colour. They can be used before or after a noun, to give more detail. *The red bus.*

adverbs can describe the manner, time, place or cause of something. They tell you more information about the event or action.

adverbials are words or phrases that give us more information about an event or action. They tell you how, when, where or why something happened.

alliteration is the repetition of a consonant sound or letter in several words: *beautiful black butterfly.*

analogy is a comparison in which an idea is compared to something that is quite different. It compares the idea to something that is familiar to the reader. *There are plenty more fish in the sea.*

antonyms are words with opposite meanings. *Hot – cold; light – dark.*

apostrophes:
- show the place of missing letters (**contraction**)
- show who or what something belongs to (**possession**).

assonance is the repetition of a vowel sound in several words: *aggressive angry alligator.*

auxiliary verbs are *be, have, do* and the **modal verbs**.

B

brackets show parenthesis. They are placed around extra information in a sentence. *Alex (who had got up late) ran all the way to school.*

C

clauses are groups of words that must contain a subject and a verb. Clauses can sometimes be complete sentences.
- **main clause:** contains a subject and verb and makes sense on its own.
- **subordinate clause:** needs the rest of the sentence to make sense. A subordinate clause includes a conjunction to link it to the main clause.
- **relative clause:** is a type of subordinate clause that changes a noun. It uses relative pronouns such as *who, which* or *that* to refer back to that noun.

colons are used to introduce a list, quotations and for separating two equal clauses.

command tells someone to do something and ends with an exclamation mark or a full stop.

commas have different uses including:
- to separate items in a list
- to separate a fronted adverbial from the rest of the sentence
- to clarify meaning
- to show parenthesis.

common noun (*boy, man*) names something in general.

conjunctions link two words, phrases or clauses together. There are two main types of conjunction:
- **co-ordinating conjunctions** (*and, but*) link two equal clauses together.
- **subordinating conjunctions** (*when, because*) link a subordinate clause to a main clause.

consonants are most of the letters of the alphabet except the vowel letters *a, e, i, o, u*

contraction a shortened word where an apostrophe shows the place of missing letters.

co-ordinating conjunctions (*and, but*) link two equal clauses together.

D

dashes in pairs show parenthesis. A single dash can also be used instead of a colon.

determiners go before a noun (or noun phrase) and show which noun you are talking about.

direct speech is what is actually spoken by someone. The actual words spoken will be enclosed in **inverted commas**: *"Please can I have a drink?"*

E

ellipsis is the omission of a word or phrase which is expected and predictable. *Where are you going? (To) the shops. Ellipsis = to.*

exclamation expresses excitement, emotion or surprise and ends with an exclamation mark.

F

figurative language uses words and ideas to create a mental image. Imagery, metaphors, similes and personification are all types of figurative language.

fronted adverbials are at the start of a sentence. They are usually followed by a comma.

future time is shown in a number of different ways. These all involve the use of a present tense verb. *We will go to the park. We are going home tomorrow.*

H

homophones are words that sound the same but are spelled differently and mean different things.

hyphens are used to join words together or to clarify meaning. A man-eating shark.

I

imagery uses words that create a picture of ideas in our minds.

inverted commas (also known as speech marks) are punctuation that enclose direct speech: "Please can I have a drink?"

M

main clause: contains a subject and verb and makes sense on its own

metaphors describe something as being something else, even though it is <u>not</u> actually that. The moon was a ghostly galleon.

modal verbs tell us how likely it is that something will happen.

N

nouns are sometimes called 'naming words' because they name people, places and things.
- **proper noun** (Ivan, Wednesday) names something specifically and starts with a capital letter.
- **common noun** (boy, man) names something in general.

noun phrases are phrases with nouns as their main word and may contain adjectives or prepositions: enormous grey elephant/in the garden.

O

object is normally a noun, pronoun or noun phrase that shows what the verb is acting upon.

P

parenthesis is a word, clause or phrase inserted into a sentence to add more detail.

passive voice is when the subject has the action performed on it. The sentence It was eaten by our dog is the passive of Our dog ate it.

past tense describes past events. Most verbs take the suffix ed to form their past tense.

perfect form of a verb usually talks about a past event and uses the verb have + another verb.
- **past perfect**: He had gone to lunch.
- **present perfect**: He has gone to lunch.

personal pronouns replace people or things.

personification is when human qualities are given to an animal, object or thing.

phrase is a group of words that are grammatically connected so that they stay together, and that expand a single word. Phrases do not contain a subject or a verb.

plural means 'more than one'.

possession a word that shows who or what something belongs to using an apostrophe.

possessive pronouns are used to show who something belongs to.

prefix is a set of letters added to the beginning of a word in order to turn it into another word.

prepositions link nouns (or pronouns or noun phrases) to other words in the sentence. Prepositions usually tell you about place, direction or time.

present tense describes actions that are happening now.

progressive or 'continuous' form of a verb describes events in progress. We are singing.

pronouns are short words used to replace nouns (or noun phrases) so that the noun does not need to be repeated.
- **personal pronouns** replace people or things.
- **possessive pronouns** are used to show who something belongs to.
- **relative pronouns** introduce a relative clause and are used to start a description about a noun.

proper noun (Ivan, Wednesday) names something specifically and starts with a capital letter.

Q

question asks a question and ends with a question mark.

R

relative clause is a type of subordinate clause that changes a noun. It uses relative pronouns such as who, which or that to refer back to that noun.

relative pronouns introduce a relative clause and are used to start a description about a noun.

root word is a word to which new words can be made by adding prefixes and suffixes: happy – unhappy – happiness.

S

semi-colons can be used to separate longer items in a list and to separate two unequal clauses.

entence is a group of words which have a subject nd verb and make sense. There are different types of entence:

- **statement** is a fact which ends with a full stop.
- **question** asks a question and ends with a question mark.
- **command** tells someone to do something and ends with an exclamation mark or a full stop.
- **exclamation** expresses excitement, emotion or surprise and ends with an exclamation mark.

imiles use words such as 'like' or 'as' to make a irect comparison.

ingular means 'only one'.

tatement is a fact which ends with a full stop.

ubject of a verb is normally the noun, noun phrase or ronoun that names the 'do-er' or 'be-er'.

ubjunctive is used in formal language. I wish I were...

ubordinate clause needs the rest of the sentence o make sense. A subordinate clause includes a onjunction to link it to the main clause.

subordinating conjunctions (when, because) link a subordinate clause to a main clause.

suffix is a word ending or a set of letters added to the end of a word to turn it into another word.

syllable sounds like a beat in a word. Longer words have more than one syllable.

synonyms have the same or a similar meaning.

T

tense is **present** or **past** tense and normally shows differences of time.

V

verbs are doing or being words. They describe what is happening in a sentence. Verbs come in different tenses.

vowel sounds are made with the letters a, e, i, o, u. Y can also represent a vowel sound.

W

word families are groups of words that are linked to each other by letter pattern or meaning.

Maths glossary

12-hour clock Uses 12 hours, with am before 12 noon, and pm after.

24-hour clock Uses 24 hours for the time; does not need am or pm, such as 17:30 = 5.30pm.

2D Two-dimensional, a term used for a shapes with no depth, usually drawn on paper.

3D Three-dimensional, a term used for solid shapes with length, depth and height.

A

Acute angle An angle between 0° and 90°.

Adjacent Near to, or next to, something. Usually used for talking about angles, sides or faces as the properties of a shape.

Algebra The use of symbols to represent numbers.

Analogue clock Shows the time with hands on a dial.

Angle The measure of the gap between lines or of rotation, measured in degrees.

Anti-clockwise Rotating in the opposite direction to the hands of a clock.

Approximate A number found by rounding or estimating.

Area The amount of surface covered by a shape.

Axis (plural axes) The horizontal and vertical lines on a graph.

B

Base 10 The structure of our number system, also called 100s, 10s, 1s and Powers of 10.

C

Circumference The edge of a circle, which is always the same distance from the centre.

Clockwise Rotating in the same direction as the hands of a clock.

Coordinates Numbers that give the position of a point on a graph, (x, y).

Cube number A number multiplied by itself twice, such as $2 \times 2 \times 2 = 2^3 = 8$.

D

Decimal places The position of numbers to the right of the decimal point. Tenths, hundredths, and so on.

Decimal point The dot used to separate the fractional part of a number from the whole.

Denominator The number on the bottom of a fraction

Diameter The maximum distance across a circle. The diameter is two times the radius.

Difference The amount between two numbers.

Digits Our number system uses ten digits, 0–9, to represent all our numbers.

Digital clock Shows time using digits rather than by having hands on a dial, often uses 24-hour time.

E

Edge The line where two faces of a 3D shape meet.

Equation A sentence of numbers, variables and operations that balances about the = sign.

Equivalent fractions Two or more fractions where the same amount is represented differently, such as $\frac{1}{2}$ and $\frac{2}{4}$.

Estimate To use information to get an approximate answer.

Even numbers Whole numbers that can be divided by 2. They end in 0, 2, 4, 6 or 8.

F

Face The flat or curved areas of 3D shapes.

Factor A number that will divide exactly into a particular number. 4 is a factor of 12.

Formula An equation used for calculating particular quantities, such as the area of a circle.

I

Imperial units Traditional units for measuring length, capacity, mass, such as pints and ounces.

Improper fraction (or vulgar fraction) A fraction with a larger numerator than denominator.

Irregular polygon A 2D shape which does not have identical sides and angles.

Isometric A drawing: a technique for drawing 3D shapes on flat surfaces.

L

Line graph A graph used to show changes over time, such as height, temperature or speed.

M

Mean The average of a set of data. The total divided by the number of items.

Mental methods Approaches for accurately solving calculations without writing them down.

Million The number 1,000,000; one thousand times one thousand.

Mixed number The combination of a whole number and a fraction, such as $3\frac{2}{5}$.

Multiple A number made by multiplying two numbers together. 6 is a multiple of 2, and also a multiple of 3.

N

Negative number A number less than zero.

Numerator The top number of a fraction. The numerator is divided by the denominator.

O

Obtuse angle An angle between 90° and 180°.

Odd numbers Whole numbers that cannot be divided exactly by 2. They end in 1, 3, 5, 7 or 9.

P

Percentage A number expressed as a fraction out of 100, such as 58%.

Perimeter The distance around the edge of a closed shape.

Pie chart Data represented as proportions of 360°, shown as fractions of a circle.

Polygon Any straight-sided 2D shape.

Positive number A number greater than zero.

Powers of ten The structure of our number system, also called Base 10 or 100s, 10s, 1s.

Prime factor A factor that is also a prime number. 3 is a prime factor of 12.

Prime number A whole number that can only be divided exactly by itself and 1.

Proportion The fraction of an amount, such as eight out of nine people wore red.

Q

Quadrants The four sections of a coordinate grid, positive and negative.

R

Radius The distance from the centre of a circle to the circumference.

Range The difference between the smallest and largest numbers in a data set.

Ratio The comparison of quantities, such as there is one black bead for every three white.

Recurring decimal A decimal that repeats the same number or numbers forever.

Reflection Changing the coordinates of a point or shape in a mirror line.

Reflex angle An angle measuring between 180° and 360°.

Regular polygon A 2D shape with all sides and angles identical.

Roman numerals The system of letters used by the Romans to represent numbers.

Rounding To simplify a number to the nearest power of 10.

S

Square number A number multiplied by itself, such as $3 \times 3 = 9$.

Symbol A sign used for an operation or relationship in mathematics, such as +, −, ×, ÷ or =, <, >.

Symmetrical A symmetrical shape is one that is identical either side of a mirror line.

T

Translation To move the coordinates of a point or shape, by the same amounts, on a graph.

V

Vertex (plural vertices) The corner of a 3D shape where edges meet.

English answers

Page 10

1   **a.** The <u>mischievous</u> toddler hid in the <u>large</u> cupboard.
    b. It was a <u>disastrous</u> start to their <u>annual</u> holiday.

2   Accept any appropriate substitution of *nice*, for example
    a. Josh wrote a **fantastic** story.
    b. Aliah enjoyed the **exciting** pantomime.

3   Accept appropriate adjectives, for example
    a. sleepy, snowy
    b. massive, vicious

Page 11

1

Word	Common noun	Proper noun
summer	✓	
Turkey		✓
pleasure	✓	

Page 12

1

Present tense	Past tense	Present progressive	Past progressive
she brings	she **brought**	**she is bringing**	she was bringing
they catch	they caught	**they are catching**	they were catching
it grows	**it grew**	it is **growing**	**it was growing**
we build	we built	**we are building**	we **were building**

2   They **were working** hard when the fire alarm stopped them.

Page 13

1   **a.** He **has gone** out to play.
    b. They **have developed** a method for baking perfect bread.

2   I **had enjoyed** the film until the end spoilt it.

Page 14

1   I **saw/see** a bird in the garden.

2   has accompanied

Page 15

1   We <u>could</u> stay in on Saturday night but we <u>might</u> go to the cinema instead.

2   George **must/should/ought to** improve his backhand if he wants to win the tennis match.

3   Emma will buy some jeans on Saturday. ✓

Page 16

1   **a.** He (gently) stroked the frightened kitten.
    b. They ran (desperately) to catch the train.

2   Accept any appropriate adverb, for example
    a. She **quickly** opened the enormous parcel.
    b. We **desperately** searched the gloomy forest.

Page 17

1

Sentence	Adverb of time	Adverb of place	Adverb of manner
Rarely has the show been so successful.	✓		
She practised **hard** for the piano test.			✓
They didn't know the treasure was **nearby**.		✓	

Page 18

1   Answers may vary, For example:
    a. It is **definitely** six miles to town.
    b. I can **possibly** come to see you later.
    c. Maybe we can have tea together?

2   Accept any appropriate explanation, for example
    Clearly suggests they expect to win.
    Possibly suggests some doubt about whether they will win.

Page 19

1   In the garden, the puppies played happily.

2   Accept any appropriate fronted adverbial, for example
    After rushing, they arrived at the party early.

Page 20

1

Group of words	Clause	Not a clause
they came home	✓	
because they		✓
it was a wonderful beach holiday	✓	

2   **a.** Despite the long delay, <u>they arrived on time</u>.
    b. <u>They studied hard</u> for their test.

Page 21

1   Accept any appropriate main clause, for example
    They took the dog for a walk in the evening.
    I love pizza, although I can't make one.

2   Accept any subordinate clause which makes sense, for example
    a. I watched television until **it was bedtime**.
    b. We haven't got much bread though **there is enough for a sandwich**.

Page 22

1   **a.** My new bike was light **so** I was able to go very fast.
    b. I like curry **but** I don't like it very spicy.
    c. I wasn't able to score a goal **nor** was I able to help my team score.

2   **a.** I wanted a new tablet **but** they were very expensive.
    b. The house was very cold **but** the central heating was on.

Page 23

1   **a.** I can't go swimming **unless** you give me a lift.
    b. I will go out with you **if/since/as** you are free.

2   Accept any appropriate subordinate clause, for example
    a. More people came in after **the play started**.
    b. Even though you are my elder sister **you are shorter than me**.

Page 24

1   Accept any appropriate relative clause, for example
    a. The hotel, **which we liked**, was next to the beach.
    b. August, **when it is school holidays**, is very busy.

Sentence	Main Clause	Subordinate clause	Relative clause
The rain, **which fell heavily**, made us cancel the trip.			✓
We called at Tomas's house **after we had seen Josh**.		✓	
Unless you are able to pay tomorrow, **the trip will be full.**	✓		

1   **a.** Alicia enjoyed the party but **she** didn't like the food.
    b. George and Oscar went sledging which **they** found enthralling.

2   **a.** I have never used my <u>fountain pen</u> as **it** is too messy!
    b. <u>John and I</u> both devoured **our** food.

Page 26

1   **a.** The hawk circled **around** its prey.
    b. He took the milk **out of** the fridge.

2   Accept any sentences which use beneath and across appropriately, for example
    He found the mushrooms **beneath** the tree.
    She kicked the football **across** the park.

Page 27

1   **a.** I washed **my** face with **some** soap.
    b. We climbed up **the** stairs and reached **our** bedrooms.

2   (Every) child must pay (some) money for (the) school trip.

Page 28

1   **a.** My (mum) drove the car.
    b. Our (cat) ate its food.

2   **a.** Dad is making (tea).
    b. The dog chased the (cat).

Page 29

1

Sentence	Active	Passive
The winning shot was made by Alisha.		✓
The team won the league.	✓	
Small mammals are hunted by eagles.		✓
Many people have climbed Mount Everest.	✓	

2   A wonderful meal was made by the chef.

Page 30

1   **a.** It is important that you **be** on time for the show.
    b. If I **were** you, I would take the risk.

2   **a.** If I <u>were</u> to give you £25, what would you do with it?
    b. The teacher asked that her students <u>be</u> quieter.

PUNCTUATION

Page 31

1   It is a sunny day. —————————→ statements
    Is it sunny? ——————⟍
    What time does it start? ——⟍⟋
    We can start it soon. ——————→ questions

2   Accept any question starting with the word 'When', for example
    When can we go to the cinema?

Page 32

1

Sentence	Statement	Question	Exclamation	Command
Do you want a new bicycle		✓		
Racing bikes are very aerodynamic	✓			
What an amazing bicycle			✓	
Ride this bike				✓

2   Accept any appropriate exclamation, starting with 'How', for example **How fantastic of you to come!**

Page 33

1   **a.** You won't be late, <u>will you</u>?
    b. We're going to the cinema, <u>aren't we</u>?

2   **a.** You'd like pizza for tea, **wouldn't you**?
    b. This is the right answer, **isn't it**?

Page 34

1   **a.** I wonder if (it'll) be sunny later.
    b. I (should've) sent a birthday card to my gran.

2   **a.** hadn't
    b. could've
    c. we'd

Page 35

1   **a.** The girls' bags
    b. The boy's crayons

2   The trains' arrivals were all delayed by the weather.

Page 36

1   **a.** We had Jack, Amir, Rashid and Josef on our team.
    b. The children enjoyed their picnic of sausage rolls, egg sandwiches, apples, crisps and juice.

2   Europe is made up of many countries including Britain, France, Spain, Germany and Italy. ✓

Page 37

1   **a.** My favourite city is Paris, which is the capital of France.
    b. Paris, which is my favourite city, is the capital of France.

Page 38

1   **a.** My mum loves cooking, my dad and me.
    b. Nate invited two boys, John and Eddy.
    c. My uncle, a singer and a dancer, often appeared on television.
    d. Has the cat eaten, Callum?

Page 39

1   **a.** Grim and sinister, the graveyard lay before me.
    b. After lunch, we had geography and art.
    c. Patiently, I waited my turn.
    d. While the lead singer sang loudly, the guitarist played the backing tune.
    e. Silently and softly, the snow fell outside.
    f. Running to catch the bus, I tripped and fell.

2   Accept any appropriate fronted adverbial, for example **Despite forgetting to buy the burgers,** it was a wonderful barbeque.

Page 41

1   **a.** "We can sit over there," said Demi.
    b. Sherri said, **"This punctuation stuff is easy."**

2   **a.** Donny said, **"I like eating cream cakes."**
    b. "They're not good for you," replied Shirley. *or* "They're not good for you!" replied Shirley.

3   **a.** My sister said, "On Sunday we are going to Nan's for tea."
    b. The driver said, "This bus is going to town."

Page 43

1   **a.** You will need to bring with you: your passport, plane tickets, money, sun cream and sunglasses.
    b. We now know some countries that border the Mediterranean Sea: Egypt, France, Spain and Italy.
    c. Warm waters can be found in: the Mediterranean Sea, the Caribbean Sea and the Indian Ocean.
    d. In *Julius Caesar*, Shakespeare wrote: "There is a tide in the affairs of men, which taken at the flood, leads on to fortune."
    e. Phoebe often wears sunglasses: the bright light hurts her eyes.

2   May has thirty days; so does June.

3 I need to go to: the supermarket for some dog food; the heel bar where my shoes have been mended; and the library to get some books for my history project.

4 Water boils at 100 degrees Celsius at sea level; it freezes at 0 degrees.

Page 45

1 **a.** The three men (they looked like spies to me) talked quietly in the corner of the cafe.

b. Denny, my older brother, is joining the army.

c. Suki – a long-haired Alsatian – won first prize at the dog show.

d. I had to keep very still while the doctor, who was very gentle, took my stitches out. (Accept dashes or brackets in place of the commas.)

e. A massive hurricane – the strongest wind ever – will hit this country, probably on Tuesday next week. *or* A massive hurricane, the strongest wind ever, will hit this country – probably on Tuesday next week.

Page 46

1 My mother-in-law is coming for Sunday lunch.

2 My uncle, a retired surgeon, showed me some of his little-used instruments.

3 re-sign

Page 47

1 **a.** My son was born <u>at the start of this century</u>, in 2001. or My son was born at the start of this century, <u>in 2001</u>.

b. I went because I wanted to <u>go</u>.

c. My sister likes salad but I don't <u>like salad</u>.

2 **a.** "**No** we don't."

b. "**Yes** I have."

c. "My dad **does**."

Page 48

1 **law.** // **Outside**. Reason: change of place.

Page 49

1 **a.** **'The way we speak'** or similar. Accept answers that summarise the entire passage.

b. **'My cousins' speech'** or similar. Accept answers that refer to differences in pronunciation.

VOCABULARY

Page 50

1 complex ✓ arduous ✓ intricate ✓

2
ancient — genuine
curious — antique
familiar — inquisitive
sincere — known

Page 51

1
healthy — minimum
young — mature
permanent — unwell
maximum — temporary

2 Accept any appropriate antonyms, for example

a. I **destroyed** a massive tower.

b. The successful man was very **arrogant**.

c. The **sensible** child had no packed lunch.

Page 52

1 **a.** **im**mature

b. **ir**relevant

c. **in**accessible

d. **il**legible

2 **a.** Emily had several (incorrect) answers.

b. They waited (impatiently) to be chosen for the team.

c. An (irregular) hexagon has six unequal sides.

Page 53

1
re — loyal   disloyal
dis — judge   misjudge
mis — design   redesign

2 replace, miscalculate, distasteful

Page 54

1 **a.** mali**cious**

b. infec**tious**

c. vigor**ous**

d. mountain**ous**

2 **a.** (religious)

b. (conscientious)

Page 55

1

Start of word	ant or ent?	ance or ence?	ancy or ency?
dec	**decent**		**decency**
confid	**confident**	**confidence**	
toler	**tolerant**	**tolerance**	**tolerancy**
obedi	**obedient**	**obedience**	

2 **a.** The (non-existence) of dodos in Mauritius has long been a cause for regret.

b. Your help is more of a (hindrance).

c. Please complete the (relevant) application form.

Page 57

1 **a.** Any two from: critical, criticise, critisim

b. Any two from: dependent, dependant, dependable, depending, depended

c. Any two from: achievable, achieving, achieved, achievement

2

Word	Root word	Suffix	Prefix
impatience	**patient**	**ence**	**im**
unfriendly	**friend**	**ly**	**un**
disappointment	**appoint**	**ment**	**dis**

3 **a.** disembark

b. indecent

c. maternal

4 **a.** enchant

b. enthral

c. apply

SPELLING

Page 58

1 **a.** They **sought** a way out of the forest, but it was hard to find.

b. I **brought** some toys with me.

c. They installed a new **wrought** iron gate.

d. We **bought** some cakes to have with our sandwiches.

2
ought — nothing
fought — considered
nought — struggled
thought — should

Page 59

1 **a.** They **roughly** worked out how to make the model.

b. The doors were made of **toughened** glass.

c. We bought some **doughnuts** to eat at the fairground.

2

Definition	Word
area	**borough**
branch	**bough**
cultivate	**plough**
sufficient	**enough**
animal food container	**trough**

Page 60

1 **a.** They went over the bridge to the Isle of Anglesey.
b. Look at the third column.
c. We are indebted to you, thanks to all your efforts.
d. Caitlin saw a Mistle Thrush in the garden.

2 **a.** (doubt)
b. (isle)
c. (condemn)
d. (bristle)

Page 61

1 **a.** I went (twice) to call on Ahmed.
b. There was a very (fierce) dog behind the gate.
c. It was (bliss) sitting in the hot sun.
d. Our teacher (suggested) that we read books by Michael Morpurgo.

2 **a.** practice
b. piece

Page 62

1

Word	1 pair	2 pairs
guarantee	✓	
accidentally		✓
pressurised	✓	
immediate	✓	

2 **a.** appreciate
b. opportunity
c. committee
d. interrupt

Page 63

1 **a.** immediately: double **m**; **iat**; is there an end **e** before **y**?
b. necessary: 1 **c** but 2 **s**; soft **s** made with **c**.

2 **a.** (government)
b. (marvellous)
c. (recognise)

Page 65

1 **a.** We heard the firework display in the park.
b. I wondered whose car was parked outside my house.
c. The burglar tried to steal the television, but it was very heavy.
d. The porridge was too hot!

2 Accept any appropriate sentence using each homophone, for example
a. My sister **passed** her driving test.
You need to meet me at half **past** three.
b. We **guessed** that you wouldn't arrive until late.
I helped mum prepare the **guest** bedroom.

3 **a.** aloud: to say something out loud
allowed: you need permission
b. farther: beyond; at a distance
father: dad
c. waste: rubbish/not making the best of
waist: middle of your body, between chest and hips

4 **a.** grate
b. descent
c. serial
d. bridle

READING

Page 66

1 what we know about the moon

Page 67

1 **a.** Last year's holiday in Menorca.
b. Any two forms of: 1. We spent a week in the pool and on the beach.
2. We never had to worry about going to bed late or getting up early.
3. Lots of new friends.

Page 69

1 **a.** Types of takeaway shops
b. Favourite takeaway meals

2 Saturday night takeaway ✓

3 Summary: My mother paints great landscapes but she is not as good at painting portraits.

Page 71

1 Mia hit the button again. Still nothing! She knew she mustn't panic. She ran down the corridor to her classroom and raced inside.
"The store room is on fire!" she shouted.
Mrs Milner took control. She told the pupils to leave everything on their desks and to go out of the building as quickly as possible. She made sure everyone had left the classroom and followed them. As she went outside she pressed the red button by the door. The fire alarm sounded.
a. The pupils are all safe outside the building. The fire brigade comes and puts out the fire.
b. These are the next logical steps from the clues in the passage.

2 **a.** He might congratulate her on saving all of the pupils and the teacher, or he might want to question her about how the fire started or how she found it.
b. Congratulate: She has saved the school. She did the correct things. She didn't panic when the fire alarm didn't work. Question: His tone was serious. She was 'interviewed'. It took place in the head teacher's office. She was alone when she found the fire. The fire alarm didn't work at first.

Page 73

1 Myth

2 **a.** Possible answers: Dorca the dragon, quest, secret of eternal dragon life, knights of Nemore, dragon magic
b. Each phrase has stereotypical elements of myths.
c. Any two from: flying dragons, knights, quests and magic.

Page 75

1 **a.** Scared, worried, apprehensive or similar.
b. 'Ella went as slowly as she could into the hall' or 'She wished she had been ill that morning'.
c. No.
d. 'She wished she had practised more' shows she has not done enough work to do well. 'The test papers were lying menacingly on the desks' suggests that she feels threatened by them.

Page 76

1 climbed

Page 77

1 hate

2 The first sentence says the writer does not like rice pudding. The last one says the writer avoids it. Hate is the only word that would fit with these two sentences.

1 **a.** a simile
   b.

Language used	Type of language and effect
Tears **burned** my eyes	This is a metaphor showing the heat of the tears.
Held me **like a nurse**	The simile shows how the mother cared for the writer until he was well again.
My father brought the **remedy – superglue**.	This is a metaphor. The superglue acts like a medicine to cure everything.

Page 81

1 **a.** Exercise can make us healthier; help us live longer; help us feel better; help us to reduce weight.
   b. Two from: walk, jog, run or swim.
   c. Cheap – indicates that exercise need not be expensive. Huge – indicates amount of difference. Both words are meant to persuade the reader to exercise.

2 **a.** 'Run!' makes the reader feel as if they are being given the instruction. The repetition emphasises how important it is to run. It makes the passage scary because you do not know why you are running.
   b. It emphasises the need to run because your life is in danger if you don't.
   c. It increases the tension because it warns you but it doesn't tell you what is there. It increases the excitement because it makes you feel that whatever is chasing you is right behind you. It makes it seem like whatever is there is too scary to look at.

Page 82

1

Feature	Feature name	Example
Language	**rhetorical question**	**Why had she hit the *send* button?**
Structural	**heading**	**Why oh why?**
Presentational	**italics**	***send***

Page 83

1 It enables readers to see and hear everything as if they were there. They are able to know what Sara is thinking.

2 It leaves the passage on a cliffhanger. The readers do not know what Lewis's text will say but they can guess from what has gone before.

Page 85

1 **a.** The public
   b. Over one hundred years
   c. It was described as unsinkable
   d. continuous

2
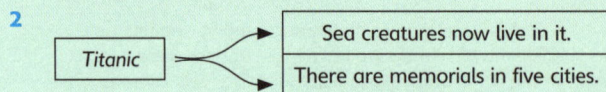

Page 87

1 **a.** One from: Owls hop and penguins walk. Penguins can swim, owls can't. Owls can fly, penguins can't.
   b. The way they move (graceful) or where they live (in the wild). Accept specific details.

2 **a.** The author would let a cat slide onto his/her knee but would not do the same with a snake.
   b. Cats and snakes both hiss and spit. They both hunt small animals.
   c. Cats are almost universally liked while snakes are hated the world over.

Page 89

1

	Fact	Opinion
The Mona Lisa is 77cm by 53cm.	✓	
The Mona Lisa is a strange picture.		✓
The Mona Lisa is not awe-inspiring.		✓
The Mona Lisa is in a dark room.	✓	

2 The Mona Lisa was painted by Leonardo da Vinci. It is also called La Gioconda. It is in the Louvre.

3 Any three from: The statue of Venus de Milo is much more impressive. It's not worth queuing to see the Mona Lisa. You'd be better off spending your time in the Egyptian section. The sphinx in there is really impressive.

Maths answers

NUMBER AND PLACE VALUE

Page 93

1. **a.** 350 **b.** 190 **c.** 3500 **d.** 1666

2. **a.** four hundred or 400 **b.** thirty thousand or 30,000
    c. four million or 4,000,000 **d.** six hundred thousand or 600,000

Brain-teaser 1,000,001
Brain-buster 9,999,999
Nine million, nine hundred and ninety-nine thousand, nine hundred and ninety-nine

Page 95

1 Eight hundred and forty-five thousand, two hundred and eighty-three

2 604,190

3 Six hundred thousand or 600,000

4 97,612 500,000 825,421 6,899,372 10,000,000

5 **a.** 3521 < 5630 **b.** 15,204 > 9798
   c. 833,521 > 795,732

Brain-teaser Madrid
Brain-buster Paris, Rome, Madrid

Page 97

1 **a.** 5000 **b.** 23,000 **c.** 45,000 **d.** 79,000

2 **a.** 100,000 **b.** 500,000 **c.** 1,400,000 **d.** 8,000,000

3 **a.** 6,000,000 **b.** 1,000,000 **c.** 4,000,000 **d.** 10,000,000

4 **a.** 0, 100,000, 200,000, 300,000, 400,000, 500,000
   b. 370,000, 380,000, 390,000, 400,000, 410,000
   c. 7,500,000, 8,500,000, 9,500,000, 10,500,000, 11,500,000

Brain-teaser

City	Rome	Paris	Madrid
Population	3,000,000	2,000,000	3,000,000

Brain-buster 8,000,000. This is lower than the actual total because there has been more rounding down than rounding up.

Page 99

1 **a.** –2 **b.** –4 **c.** 3 **d.** 0

2 –20 –16 –12 –8 –4 0 4 8 12 16 20

3 **a.** – **b.** + **c.** – **d.** –

4 **a.** 14 **b.** 19 **c.** 7 **d.** –9

Brain-teaser 8°C
Brain-buster 69.4°C

CALCULATIONS

Page 101

1 **a.** 792 **b.** 5526 **c.** 479,369

2 **a.** 540 **b.** 117,450 **c.** 2355

3 **a.** 548,704 **b.** 962,825 **c.** 5,167,467

4 **a.** 79,740 **b.** 635,231 **c.** 2,482,597

Brain-teaser 982,136
Brain-buster 8,312,272

Page 103

1 **a.** 4800 **b.** 62,000 **c.** 1600 **d.** 50,000 **e.** 430,000
   f. 1,000,000

2 **a.** 2000 **b.** 25 **c.** 40,000 **d.** 90,000 **e.** 80,001
   f. 25,000

3 **a.** 27,072 **b.** 723

Brain-teaser £160,000
Brain-buster 3000 tickets

Page 105

1 **a.** 868 **b.** 7150 **c.** 13,770 **d.** 329,576

2 **a.** 8925 **b.** 38,010 **c.** 79,890 **d.** 567,840

Brain-teaser 24,984 (!)
Brain-buster £729,723

Page 107

1 **a.** 23 **b.** 24 r3 **c.** 434 r1 **d.** 313 r2

2 **a.** 12 r2 **b.** 64 r2 **c.** 460 r5 **d.** 1132 r4

Brain-teaser 13 each with 2 stickers left over
Brain-buster 1248 tickets
You can check your answer by multiplying the number of tickets by the ticket price.

Page 109

1 **a.** 210 r14 **b.** 254 r8

2 **a.** 22 r8 **b.** 211 r7 **c.** 353 r4 **d.** 228 r22

Brain-teaser 134 rows
Brain-buster £2341.75

Page 111

1 **a.** 0 **b.** 6 **c.** 27

2 **a.** 2 **b.** 12 **c.** 5

3 **a.** correct **b.** correct **c.** incorrect (–7) **d.** correct

4 **a.** 8 × (5 + 2) – 3 = 53 **b.** 14 ÷ 7 + 2 × (11 – 6) = 12
   c. 64 – (12 + 5 × 3) = 37

Brain-teaser Yes. (34 + 17 + 43) × 2 – 20 = 168
Could also be expressed as (34 x 2) + (17 x 2) + (43 x 2) –20 = 168
Brain-buster 12,000 × 2 + (7000 – 2500) × 3 = £37,500
2 new cars and 3 second hand cars.

Page 113

1 1, 2 and 4

2 1, 2, 5 and 10

3 15, 30, 45, 60, 75, 90, etc.

4 2 × 5 × 7 = 70

5 2, 3 and 5

6 94: 2 × 47

Brain-teaser 38 has prime factors, but it is not a prime number. Prime numbers only have themselves and 1 as factors.
Brain-buster 6

FRACTIONS, DECIMALS AND PERCENTAGES

Page 115

1 **a.** 2 **b.** 3 **c.** 1 **d.** 20 **e.** 30 **f.** 11

2 **a.** True **b.** True **c.** False **d.** True

3 **a.** $\frac{3}{4}$ **b.** $\frac{3}{4}$ **c.** $\frac{3}{4}$ **d.** $\frac{3}{4}$ **e.** $\frac{9}{20}$ **f.** $\frac{5}{8}$ **g.** $\frac{32}{75}$ **h.** $\frac{8}{25}$

Brain-teaser $\frac{8}{25}$

Brain-buster $\frac{17}{25}$

Page 117

1 **a.** $\frac{15}{30}$ **b.** $\frac{20}{30}$ **c.** $\frac{18}{30}$ **d.** $\frac{25}{30}$

2 **a.** = **b.** > **c.** > **d.** <

3 **a.** True **b.** True **c.** False

4 **a.** $\frac{5}{8} < \frac{2}{3} < \frac{3}{4}$ **b.** $\frac{1}{3} < \frac{3}{7} < \frac{4}{9}$ **c.** $\frac{13}{24} < \frac{5}{9} < \frac{7}{12}$

Brain-teaser $\frac{3}{8}$ ($\frac{3}{8} = \frac{15}{40}$ and $\frac{7}{20} = \frac{14}{40}$)

Brain-buster cats ($\frac{21}{84}$) or $\frac{3}{12}$) < dogs ($\frac{24}{84}$ or $\frac{2}{7}$)) < no pets ($\frac{39}{84}$ or $\frac{13}{28}$)

Page 119

1   **a.** $\frac{5}{6}$ **b.** $\frac{7}{10}$ **c.** $\frac{7}{8}$

2   **a.** $\frac{1}{8}$ **b.** $\frac{4}{9}$ **c.** $\frac{11}{60}$

3   **a.** +   **b.** –   **c.** +   **d.** –   **e.** –   **f.** +

4   **a.** $4\frac{1}{4}$ **b.** $1\frac{1}{4}$ **c.** $1\frac{2}{15}$ **d.** $4\frac{7}{15}$

Brain-teaser $\frac{1}{6}$

Brain-buster $\frac{16}{77}$

Page 121

1   **a.** 10   **b.** 6   **c.** 18   **d.** 10   **e.** 25   **f.** 26

2   **a.** $3\frac{1}{2}$ **b.** $12\frac{1}{2}$ **c.** $13\frac{1}{3}$ **d.** 6   **e.** $7\frac{1}{5}$ **f.** $16\frac{2}{3}$

3   **a.** $\frac{1}{6}$ **b.** $\frac{6}{20}$ or $\frac{3}{10}$ **c.** $\frac{24}{72}$ or $\frac{1}{3}$ **d.** $\frac{20}{30}$ or $\frac{2}{3}$ **e.** $\frac{10}{24}$ or $\frac{5}{12}$

   f. $\frac{40}{35}$ or $1\frac{1}{7}$

Brain-teaser $5\frac{1}{4}$ minutes (or 5 minutes 15 seconds)

Brain-buster $\frac{1}{3600}$

Page 123

1   **a.** right   **b.** right   **c.** wrong   **d.** right   **e.** wrong   **f.** right

2   **a.** $\frac{1}{4}$ **b.** $\frac{1}{12}$ **c.** $\frac{1}{15}$ **d.** $\frac{1}{6}$ **e.** $\frac{3}{16}$ **f.** $\frac{1}{30}$

Brain-teaser $\frac{1}{14}$ of the whole cake

Brain-buster $\frac{1}{80}$ of the sheet; 3 stickers per child.

Page 125

1   **a.** 0.4   **b.** 0.6   **c.** 0.375

2

Fraction	$\frac{1}{8}$	$\frac{2}{8}$	$\frac{3}{8}$	$\frac{4}{8}$	$\frac{5}{8}$	$\frac{6}{8}$	$\frac{7}{8}$	$\frac{8}{8}$
Decimal	0.125	0.25	0.375	0.5	0.625	0.75	0.875	1 or 1.0

3   $\frac{3}{4} = 0.75$, $\frac{5}{8} = 0.625$, $\frac{4}{5} = 0.8$, $\frac{1}{3} = 0.333$

4   $0.166 = \frac{1}{6}$, $0.4 = \frac{2}{5}$, $0.7 = \frac{7}{10}$, $0.125 = \frac{1}{8}$

Brain-teaser $\frac{5}{6}$ (=0.833)

Brain-buster He is wrong. $\frac{1}{12} = 0.083$ and $\frac{1}{10} = 0.1$

Page 127

1   **a.** 0.375: 5 thousandths, 7 hundredths, 3 tenths
    b. 0.903: 3 thousandths, 0 hundredths, 9 tenths

2

Fraction	Decimal	3dps	2dps	1 dps
$\frac{2}{7}$	0.285714	0.286	0.29	0.3
$\frac{3}{13}$	0.230769	0.231	0.23	0.2
$\frac{4}{11}$	0.363636	0.364	0.36	0.4
$\frac{2}{3}$	0.666666	0.667	0.67	0.7
$\frac{8}{9}$	0.888888	0.889	0.89	0.9

Brain-teaser Jared is wrong. It would be rounded down to zero point zero.

Brain-teaser It is a recurring number because 3 ÷ 11 is 0.272727. Rounded to 3dp it is 0.273.

Page 129

1   **a.** 0.6   **b.** 6.6   **c.** 0.92   **d.** 2.04

2   **a.** 4.83   **b.** 6.75   **c.** 6.25   **d.** 109.89

Brain-teaser £9.20
Brain-buster £15.20

Page 130

1   **a.** 0.13   **b.** 0.27   **c.** 0.24

2   **a.** 0.04   **b.** 0.22   **c.** 0.15   **d.** 5.16

Brain-teaser £3.48

Page 131

1

Percentage	Decimal	Fraction
33.3%	0.333	$\frac{1}{3}$
12.5%	0.125	$\frac{1}{8}$
40%	0.4	$\frac{2}{5}$
85%	0.85	$\frac{17}{20}$
87.5%	0.875	$\frac{7}{8}$

Brain-teaser 40%

RATIO AND PROPORTION

Page 133

1   **a.** 4 in 9 or $\frac{4}{9}$ **b.** 1 in 2 or $\frac{1}{2}$ **c.** 1 in 5 or $\frac{1}{5}$

2   **a.** 1:2   **b.** 1:2   **c.** 3:4

Brain-teaser a. 1 in 5 can speak two languages
b. dual to single = 1:4
Brain-buster a. 20 blueberries
b. The proportion of blueberries will be 4 in 7

Page 135

1   **a.** 25%   **b.** 70%   **c.** 40%   **d.** 37.5%

2   **a.** 1 in 4   **b.** 2 in 5   **c.** 13 in 50   **d.** 7 in 8

3   **a.** 50   **b.** 0.5   **c.** 62.4   **d.** 285   **e.** 14.4   **f.** 54

4   See glossary on pages 94 and 95

Brain-teaser 55 cars
Brain-buster 7,560,000 dogs

Page 137

1   **a.** 8cm   **b.** 20cm   **c.** 40cm

2

Scale	Side length	Area
5:1	5cm	25cm²
10:1	10cm	100cm²
25:1	25cm	625cm²

3   **a.** 50cm   **b.** 20cm   **c.** 5cm

Brain-teaser 5.5m or 550cm
Brain-buster 7.5:1

ALGEBRA

Page 139

1

Length	Width	P	A
5cm	2cm	14cm	10cm²
5m	4m	18m	20m²
7km	1.5km	17km	10.5km²
3.2m	2.3m	11m	7.36m²

2

h	11	14	20	35	308
f	1	2	4	9	100

Brain-teaser $334
Brain-buster

Fahrenheit	32°	104°	212°
Celsius	0°	40°	100°

Page 141

1 **a.** 8 **b.** 18 **c.** 25 **d.** −2 **e.** 6 **f.** 7 **g.** 4 **h.** 33

2 **a.** 62 **b.** 6 **c.** 10 **d.** −3

Brain-teaser $n = £42.50 − (25 × £1.50)$
$n = £5$
Brain-buster $n = (190.40 ÷ 2) ÷ 5.60$
$n = 17$ children

Page 143

1 **a.**

x	0	1	2	3	4	5	−1	−2	−3
y	2	3	4	5	6	7	1	0	−1

b.

s	0	2	5	6	7	8	9	10	−1
t	8	6	3	2	1	0	−1	−2	9

c.

q	0	1	1.5	2	5	10	100	−1	−10
p	−3	−1	0	1	7	17	197	−5	−23

Brain-teaser a. 60 **b.** 85
Brain-buster $x = 6, y = 2$

MEASUREMENT

Page 145

1 **a.** 300 minutes **b.** 7200 seconds

c. $8\frac{1}{2}$ minutes or 8.5 minutes **d.** 86,400 seconds

2 **a.** 23,000mm **b.** 2400m **c.** 100,000cm **d.** 0.685m

3 **a.** 0.75kg **b.** 32,500g **c.** 0.001kg **d.** 350g

4 **a.** 2500ml **b.** 0.75l **c.** 63.425l **d.** 250ml

Brain-teaser 242 metres 63 centimetres and 7 millimetres
Brain-buster 31,622,400 seconds

Page 147

Brain-teaser 18.75 seconds 0.79kg or 790g 233g
Brain-buster Height = 1m 49.86cm Weight = 39.498kg
29,220 days 701,280 hours 42,076,800 minutes

Page 149

1 **a.** P = 13cm A = 9cm² **b.** P = 6cm A = 2.25cm²

2 **a.** P = 50m A = 87.5m² **b.** P = 24m A = 13.75m²

Brain-teaser 15m²
Brain-buster 180 tiles

Page 151

1 **a.** 10cm² **b.** 17.5cm² **c.** 22.5cm²

2 **a.** square **b.** triangle **c.** triangle

Brain-teaser £511.98
Brain-buster 7.875m²

Page 153

1 **a.** and **b.** Check that children's drawings are accurate.

2 **a.** 216cm³ **b.** 36m³ **c.** 1000m³ **d.** 90cm³
e. 1728mm³ **f.** 9000mm³

3 1,000,000,000 (one billion)

Brain-teaser 0.125m³, 125,000cm³
Brain-buster 42m³

GEOMETRY

Page 155

1 **a.** right angle **b.** acute angle **c.** obtuse angle

2 **a.** 90° **b.** 143° **c.** 25°

Brain-teaser All acute angles should be 55°;
all obtuse angles 125°
Brain-buster Look for understanding that parallel lines have similar angles when intersected and that angles on a straight line add up to 180. The formula might be along the lines of $2a + 2b = 360°$

Page 157

1 A regular polygon has all sides and angles equal.

2 **a.** equilateral triangle – regular
b. rhombus (quadrilateral) – irregular
c. pentagon – irregular **d.** square – regular
e. hexagon – regular **f.** heptagon – irregular

Brain-teaser The distance from the centre to each corner is identical, and the four angles at the centre are all 90°.
Brain-buster She is right. The angle at the centre must be 60° for each triangle, and the side lengths from the centre must be identical, therefore the other angles must also be 60°, and the other side an identical length.

Page 159

1 Check that all sides and angles are the same.

2 Check that angle is 120° and all sides are 3cm.

3 Construct an isosceles triangle with single angle = 360 ÷ 8 = 45°; repeat this 8 times, with the 45° angles forming a complete turn in the centre of the octagon.

Brain-teaser 1440°
Brain-buster $a = (n × 180 − 360) ÷ n$

Page 161

1 Check that all sides are connected by one tab.

2 Check that net would fold and glue correctly.

Brain-teaser Check that instructions show understanding of faces joining and dimensions of sides being correct.
Brain-buster a. 5cm × 5cm × 20cm **b.** 150cm² wasted
c. 500cm³

Page 163

1 **a.** radius: the distance from the centre of a circle to the circumference
b. diameter: the distance across the widest part of a circle, twice the radius
c. circumference: the distance around the edge of a circle.

2 7m

3 0.75km or 750m

Brain-teaser No. The circumference of a circle is in fact 3.14d, in this case = 12.56cm
Brain-buster Yes. The area of a circle is in fact 3.14r², which in this case = 12.56cm²

1 **a.** A (3, 2) B (−5, 2) C (−5, −6) D (3, −6) **b.** Square
   c. (−1, −2)

2 **a.** P (3, 5) Q (−1, 5) R (−4, −1) S (0, −1) **b.** Parallelogram

Brain-teaser A kite
Brain-buster In no particular order, (−3, 6), 2nd quadrant;
(−3, −4), 3rd quadrant; (7, −4), 4th quadrant.

Page 167

1 **a.** P'(−9, 0) Q'(−4, 0) R'(−9, −5) S'(−4, −5)
   b. W'(−3, 7) X'(−9, 7) Y'(−9, 4) Z'(−3, 4)
   c. W'(3, −7) X'(9, −7) Y'(9, −4) Z'(3, −4)

2 A'(6, −2) B'(0, −5) C'(−1, 3)

3 D'(−2, −4) E'(1, −7) F'(−3, −6)

Brain-teaser The reflected square sits on top of the original
square, with the coordinates for P and S swapped, and Q and R
swapped.
Brain-buster Sort of(!) The square will end up in the same place
by reflection or translation, but the vertices will have changed
positions.

STATISTICS

Page 169

1

Cats	Guinea pigs	Dogs	Horses	Hamsters
24	12	6	3	3

2 Use a protractor to check the angles on the pie chart.
   (1p = 1°)

Mum	Dad	Paul	Lizzie	Mary
180°	5°	45°	120°	10°

Brain-teaser Around 40°

Brain-buster Accept answers + or − 0.3 billion.

Asia	Africa	Europe	Oceania	North America	South America
4.2 billion	1 billion	0.7 billion	0.04 billion	0.6 billion	0.5 billion

Page 171

1 **a.** The walkers are not moving. **b.** 5km **c.** 7.5km/h
   d.

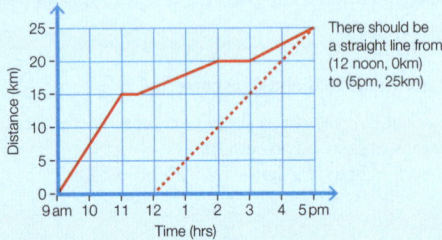

There should be
a straight line from
(12 noon, 0km)
to (5pm, 25km)

   e. 5km/h

Brain-teaser 212°F
Brain-buster x-axis: 32°F, y-axis: −18°C (answers may not be
exact due to the scale of the graph.)

Page 173

1 8

2 **a.** 145 **b.** 24.17

Brain-teaser 56
Brain-buster **a.** average = 274.5 words per page **b.** 103,212

English progress tracker

Vocabulary

Practised Achieved

Practised	Achieved		
☐	☐	Synonyms	50
☐	☐	Antonyms	51
☐	☐	Prefixes: in or im? il or ir?	52
☐	☐	Prefixes: re, dis or mis?	53
☐	☐	Suffixes: ous, cious or tious?	54
☐	☐	Suffixes: ant or ent? ance or ence? ancy or ency?	55
☐	☐	Word families	56

Spelling

Practised Achieved

Practised	Achieved		
☐	☐	Letter strings: ought	58
☐	☐	Letter strings: ough	59
☐	☐	Silent letters	60
☐	☐	c or s?	61
☐	☐	Double trouble	62
☐	☐	Tricky words	63
☐	☐	Homophones	64

Reading

Practised Achieved

Practised	Achieved		
☐	☐	Identifying main ideas	66
☐	☐	Identifying key details	67
☐	☐	Summarising main ideas	68
☐	☐	Predicting what might happen	70
☐	☐	Themes and conventions	72
☐	☐	Explaining and justifying inferences	74
☐	☐	Words in context	76
☐	☐	Exploring words in context	77
☐	☐	Enhancing meaning: figurative language	78
☐	☐	How writers use language	80
☐	☐	Features of text	82
☐	☐	Text features contributing to meaning	83
☐	☐	Retrieving and recording information	84
☐	☐	Making comparisons	86
☐	☐	Fact and opinion	88

Maths progress tracker

Number and place value

Calculations

Fractions, decimals and percentages

Ratio and proportion

Algebra

Measurement

Geometry

Statistics